Atomic Absorption Spectroscopy

Atomic Absorption Spectroscopy

A Practical Guide

R. J. Reynolds, B.Sc., A.R.C.S.
Chief Chemist, Evans Electroselenium Ltd

and K. Aldous, B.Sc., A.R.C.S.

with a chapter by

K. C. Thompson, Ph.D., B.Sc., A.R.C.S.
Senior Development Chemist, Southern Analytical Ltd
Formerly Lecturer in Chemistry, Imperial College of Science and Technology, London

Griffin 1820 London

First published . . . 1970

9×6 in., ix+201 pages
35 line and 11 tone illustrations

ISBN 0 85264 184 2

Set and printed in Great Britain by
J. W. Arrowsmith Limited, Bristol

Preface

SINCE standard commercial equipment became available, atomic absorption spectroscopy has grown more rapidly than any previous instrumental analytical technique. It has passed through the stages of initial development and application in research, and is now increasingly regarded as an essential tool in the routine laboratory. It is replacing many traditional wet methods for the estimation of metals in solution, and revealing errors in hitherto accepted procedures.

Because it does not demand intricate sample preparation, A A S is an ideal tool for the non–chemist, e.g. the engineer, biologist or clinician interested only in the significance of the results.

To save time and money for the investigator is the authors' object in this practical working guide to a powerful and versatile technique. Because it is unnecessary to comprehend an advanced mathematical philosophy to utilize the method, Theory is placed at the end of the book.

The first three chapters deal with the fundamental principles of atomic absorption, procedural considerations (interferences, selection of wavelength, flame system, sensitivities, etc.) and the general techniques of measurement. Chapters IV and V provide reference data on characteristics of the elements and the application of the technique to specific fields of analysis.

All the information necessary to develop and perform reliable analyses of metallic species in solution is therefore contained in these first five chapters.

Readers who wish to go more deeply into the subject will find information on the characteristics of commercial equipment, and on some of the techniques and hardware that have been examined in research laboratories and which may or may not be incorporated into such units in the future, in Chapters VI and VII.

We wish to acknowledge the encouragement and generous assistance given to us by Evans Electroselenium Limited. We also sincerely appreciate the assistance of Dr. K. C. Thompson who wrote the final chapter on Theory, and whose suggestions and corrections to the main text were invaluable.

Braintree, England
Summer 1970

R. J. REYNOLDS
K. ALDOUS

Contents

I

Fundamentals

Introduction

SINCE its introduction, by ALAN WALSH[1] in the mid-1950's, atomic absorption spectroscopy has proved itself to be the most powerful instrumental technique (for the estimation of metals in solution) ever presented to the analytical chemist.

Its versatility is dramatically illustrated by the fact that it allows between 60 and 70 metallic elements to be determined, in concentrations that range from trace to macro quantities[2]. The technique is not restricted to aqueous solutions, because organic and mixed aqueous–organic solvents are suitable, and in many cases advantageous, for determinations.

Prior chemical separations of constituents seldom need to be made, so that measurement of the concentration of a metallic species by atomic absorption spectroscopy is rapidly and simply accomplished. The elimination of preliminary chemical steps, such as separation of constituents, must, of course, improve the overall reliability of an analytical procedure.

The superiority of an instrumental procedure over classical wet analytical methods is becoming increasingly accepted. If such a procedure is to be of real value in the hands of an operator, especially if not specifically trained in the disciplines of analytical chemistry and instrument engineering, it must be accurate, reproducible, and rapid. Both the necessary sample preparation and the instrumental operations must be simple and economic.

In addition to the technique fulfilling these general requirements, the instrument itself should be versatile, robust, and if possible inexpensive to install. Atomic absorption spectroscopy has proved itself to come closer to the fulfilment of these desirable features than any previous physio-chemical method.

Basic Principles

If a solution containing a metallic species is aspirated into a flame (such as the air--acetylene system) an atomic vapour of the metal will usually be formed. Some of the metal atoms may be raised to an energy level sufficiently high to emit the characteristic radiation of that metal, a phenomenon that is utilised in the familiar technique of emission flame photometry. An overwhelmingly larger percentage of the metal atoms, though, will remain in the non-emitting, ground state. These ground-state atoms of a particular element are receptive of light radiation of their own specific resonance wavelength (in general, the same wavelength as they would emit if excited). Thus, if light of this wavelength is passed through a flame containing atoms of the element, part of that light will be absorbed, and the absorption will be proportional to the density of the atoms in the flame. This phenomenon is exploited in atomic absorption spectroscopy, which thereby acquires the following advantages over emission techniques.

1. The technique is intrinsically specific, since the atoms of a particular element can only absorb radiation of their own characteristic wavelength. Conversely, of course, light of a particular frequency can only be absorbed by the specific element to which it is characteristic. Spectral interferences, so troublesome in emission methods, therefore rarely occur.
2. Because of the much larger number of metal atoms that contribute to an atomic absorption signal, variation in flame temperature has relatively less effect than it does on the smaller number of atoms producing an emission signal.

Basic Instrument Design

An atomic absorption spectrophotometer consists essentially of the following components, see Fig. 1.1:

1. A stable light source, emitting the sharp resonance line of the element to be determined.

2. A flame system into which the sample solution may be aspirated at a steady rate, and which is of sufficient temperature to produce an atomic vapour of the required species from the compounds present in the solution.
3. A monochromator to isolate the resonance line, and focus it upon a photomultiplier.
4. A photomultiplier that detects the intensity of light energy falling upon it, and which is followed by facilities for amplification and readout.

The light source is usually a lamp having a hollow cathode made of the element to be determined. The emission from this lamp is modulated so that its radiation only, and not that emitted from the flame, will be recorded in the galvanometer signal. The most commonly used flame system is air–acetylene.

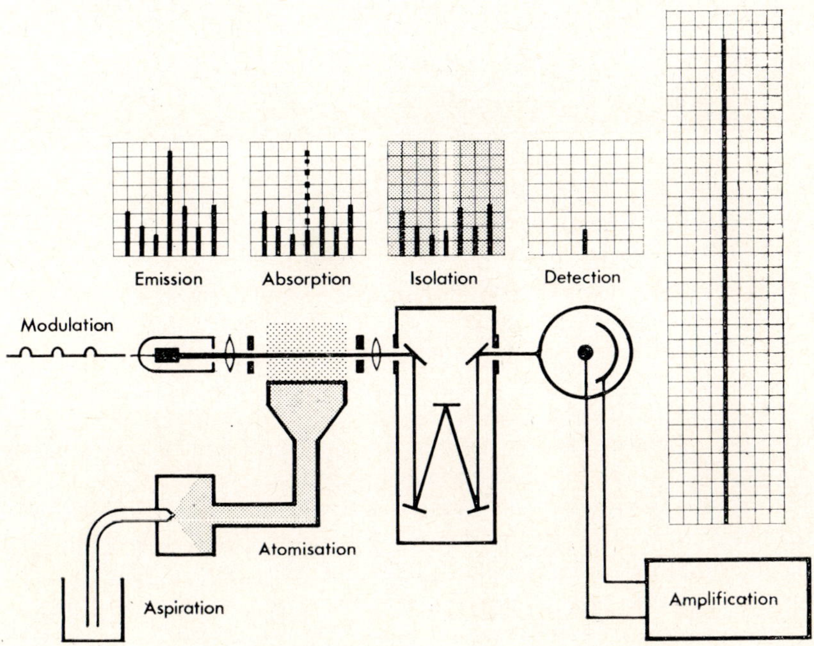

Fig. 1.1 Basic principle of spectrophotometer

Operation

In operation, a meter is adjusted to read zero absorbance when a blank solution is sprayed to the flame, and the unobstructed light of the hollow-cathode lamp passes on to the photomultiplier. When a solution containing the absorbing species is introduced, part of this light is absorbed, resulting in a diminution of light intensity falling upon the photomultiplier and giving rise to a deflection of the meter needle.

Standard solutions of the element to be determined are employed to construct a calibration curve from which the contents of test solutions can be obtained.

Advantages

Because of its freedom from spectral interferences, prior chemical separation of constituents is not required and preparation is limited to placing the sample in solution, or perhaps extracting the required species by simple elution. The very high degree of reproducibility that is easily attainable by this means is of enormous importance to the analyst. This feature of the technique is conveniently illustrated by comparing it with the most commonly used general procedure in analytical chemistry, colorimetry.

Atomic absorption is undoubtedly revealing errors due to erratic determinations in hitherto trusted procedures. The reasons why these errors can, and do, occur in colorimetry, and why atomic absorption spectroscopy is less susceptible to them, become apparent only upon considering how these procedures are practised in the routine laboratory.

The perceptive and honest analyst will acknowledge that the conditions under which a colorimetric procedure is developed and assessed are usually very different from those under which it will be used in routine practice—particularly after some time has elapsed since its introduction, with perhaps a number of junior staff changes. When carried out to perfection with no mechanical loss of the element to be determined, a colour-change method must always be liable to a positive error. Where final judgement of the change is made with the human eye, this error will be positive and variable. Visual colour matching will also be liable to some error of variation, in this case both positive and negative. Such methods usually entail meticulous attention to detail, extending in some cases even to the time allowed to elapse between addition of reagents and measurement of colour intensity. They frequently involve separations of the species to be determined, with risk of errors through losses brought about by multiple operations.

The errors due to the intrinsic positive bias of a colour-change procedure and those due to mechanical losses tend to balance each other out, and have possibly deluded analysts as to the reliability of some methods. Because of the simplicity of sample preparation required, atomic absorption is much less liable to these faults. When errors do occur they are usually glaringly large, and cause the analyst to check the instrument's operation and his standard and test solutions—actions that can usually be carried out quickly.

Chap. I References

1. WALSH, A., The application of atomic absorption spectra to chemical analysis. *Spectrochim. Acta* 7 (1955) 108.
2. GATEHOUSE, B. M., and WILLIS, J. B., Performance of a simple atomic absorption spectrophotometer. *Spectrochim. Acta* 17 (1961) 710.

II

Basic Procedural Considerations

Interferences

IT has been stated that atomic absorption spectroscopy is virtually free from spectral interferences, because a particular element can only absorb light of its own characteristic frequency; and conversely light of a particular frequency can only be absorbed by atoms of a specific element. Interferences are therefore confined mainly to phenomena that affect the number of atoms in the flame and they may be listed under the following headings.

(a) Enhancements due to higher uptake of the solution medium, as for example when an organic or mixed organic–aqueous solvent is used in place of water.

(b) Depressions due to the formation of refractory compounds that are not dissociated in the flame. The most common example of this type of interference is furnished by the well-known phosphate depression of calcium absorption when the air–acetylene flame is used.

(c) Depressions due to ionisation, as for example when a solution containing calcium is aspirated to the nitrous oxide–acetylene flame. At the temperature of this flame calcium compounds are not only dissociated to produce calcium atoms, but some of these atoms are raised to an energy level sufficiently high to produce ionisation. The calcium ions so formed will not absorb the characteristic radiation of the ground-state atom and a depression in absorption is noted.

Common elements particularly susceptible to this effect are barium, calcium, strontium, sodium, and potassium.

(d) 'Matrix' interferences can also occur. These are due to the surface tension and viscosity of a test solution being different from that of the standards, usually because of the presence of heavy concentrations of foreign ions in the former. This results in the uptake rate of the sample solution being lower than that for the standards, so that a smaller number of absorbing species per unit time are carried to the flame, and the observed absorption is therefore low.

Interferences of these types present little difficulty to the analyst. Enhancing effects may, of course, be exploited and depressive interferences are either compensated for by the use of matching standards, or overcome by the addition of suppressants.

LIGHT SCATTERING

This effect manifests itself as an apparent enhancement of the reading and is most severe at lower wavelengths. It is attributed to light being scattered from small particles of a foreign constituent that are present in the sample being analysed, but this explanation is not universally accepted[1,2]. Thc best-known example of this interference occurs with the direct estimation of lead in urine at the 2170 Å[3] line. Light of this wavelength is not only absorbed by the lead in solution, but some appears also to be scattered by the sodium chloride etc. present in heavy quantities in the urine; erroneously high readings are thereby obtained. Correction for the effect may be made either by matching the blank and standards to the approximate composition of the sample, or by noting the apparent absorption at a non-absorbing line close to the analytical wavelength and subtracting this value from the absorption at the latter.

To summarize, light scattering is of significance at wavelengths below about 2500 Å (more especially below 2200 Å) for trace determinations of metals in media that contain heavy loadings of foreign ions.

Control of Parameters that influence a Determination

The parameters that lie within the analyst's control are flame conditions, constitution of solution and, in some cases, choice of wavelength.

THE FLAME SYSTEM

Optimum and uniform flame conditions are normally attained by selecting a burner height, setting the oxidising gas (air or nitrous oxide) to a fixed flow rate, and then regulating the fuel so that peak absorption is obtained when a suitable standard solution is aspirated into the flame. If this procedure is used it is found that for most elements there is a very wide range of burner heights over which the maximum sensitivity attainable is sensibly identical.

As mentioned previously, the most commonly used flame is the air–acetylene system, and the fuel requirements for different elements vary. Some elements, such as calcium and magnesium require a flame that is rich in its fuel content, and with these elements sharply defined conditions for maximum absorption are obtained. Maximum absorption for chromium is obtained with a flame that is just short of luminous, while molybdenum requires one that is actually yellow.

Maximum absorption for a large number of elements is generally stated to be attained under fuel-lean conditions. With elements exhibiting this feature the absorption can often apparently increase up to the point where the flame eventually lifts off the burner. In many such cases, investigation in our own laboratory (EEL) has shown that the apparent increase in absorption is largely due to the flame gases themselves absorbing more and more light as the fuel supply is diminished, indicated by the "zero" reading for a blank solution following the absorption for a standard solution.

Thus by adjusting to maximum absorption by the general procedure described above a 'false' absorption peak can be obtained. Operating under such conditions a higher level of noise and a lower sensitivity are attained than with a flame having a higher fuel content.

For the elements cobalt, copper, gold, iron, lead (particularly at the 2170 Å line), nickel (particularly at the 2320 Å line) and zinc, it is advantageous to use the following procedure when adjusting the flame conditions for optimum absorption:

(a) Light the flame, set the oxidising gas flow and burner height and adjust to the correct wavelength for the element to be determined.

(b) Aspirate a blank solution and adjust the fuel gas flow so as to obtain maximum throughput of light on to the photo-multiplier (i.e. maximum transmission).

The flame conditions thus established are then used for the estimation.

In general, it can be said that the hotter the flame used, the less prone will a determination be to depression caused by chemical interferences. For example, the well-known depression caused by phosphate upon the absorption of calcium in the air–acetylene flame can be overcome by using nitrous oxide–acetylene.

The use of a hotter flame, though, can itself give rise to a depressive effect with elements such as calcium, due to their pronounced tendency at high temperatures to form ions which, of course, will not absorb the radiation from the lamp. In the case of calcium, this type of depression can be overcome by dosing the sample with a massive excess of potassium chloride. Potassium is even more readily converted to the ionic state than calcium and in doing so gives rise in the flame to a dense population of electrons which annul the tendency of calcium to ionise. The absorption thereby increases dramatically, and the benefits of high response and freedom from phosphate interference are obtained.

The absorption of chromium, although lower in the hotter nitrous oxide–acetylene flame than in the air–acetylene flame, is free from the troublesome depressive effect that the presence of iron exerts in the cooler system. Advantage can be taken of this fact to simplify the determination of chromium in steels. Molybdenum exhibits a much more linear response in the nitrous oxide–acetylene flame than with the air–acetylene system, and the sensitivity and stability are very similar.

Tungsten and titanium are remarkable in requiring a nearly luminous nitrous oxide–acetylene flame in order to obtain peak absorption. At the wavelengths used, the noise levels increase as the fuel content of the flame is raised, so that with these elements it is necessary to select flame conditions that give rise to an acceptable level of noise, which will often entail working somewhat below maximum sensitivity.

Bismuth, with absorbing lines at 2231 Å and 3068 Å, presents a unique case. The absorption at 2231 Å is more powerful than at 3068 Å, but at concentrations above about 75 μg/ml it drops off badly, so that it becomes advantageous to use the higher wavelength for determinations on solutions containing the element over widely varying concentrations. Unfortunately, though, at 3068 Å very intense hydroxyl emission occurs when the air–acetylene flame is used, but this effect is virtually absent with the cooler air–propane system. The sensitivity with this latter flame is identical to that obtainable with air–acetylene.

The response for tin is more sensitive and stable in the air–hydrogen flame than in the nitrous oxide–acetylene or luminous air–acetylene systems, but unfortunately severe interferences are observed with the cooler air–hydrogen flame, which preclude its use for some applications.

Wavelength

Choice of wavelength can be an important factor in the success or otherwise of a determination. For most purposes this is limited to mere selection of the most absorbing line, though cases do occur in which the decision requires more consideration.

The flame systems themselves emit light in different parts of the spectrum. The air-acetylene flame exhibits a series of peaks between 3200 Å and 4200 Å while with the nitrous oxide-acetylene system the greatest emission occurs between 3500 and 4400 Å as shown in Fig. 2.1 and Fig. 2.2.

The emission manifests itself mainly as background noise and can affect the choice of wavelength for a particular determination. Thus the choice of wavelength is often influenced by the flame system selected, which itself frequently depends upon possible interfering substances in the solution to be analysed. As mentioned earlier, for the determination of chromium in steels it is advantageous to use the nitrous oxide–acetylene flame in order to avoid the troublesome interference from iron that occurs when the cooler air–acetylene flame is used. With the hotter

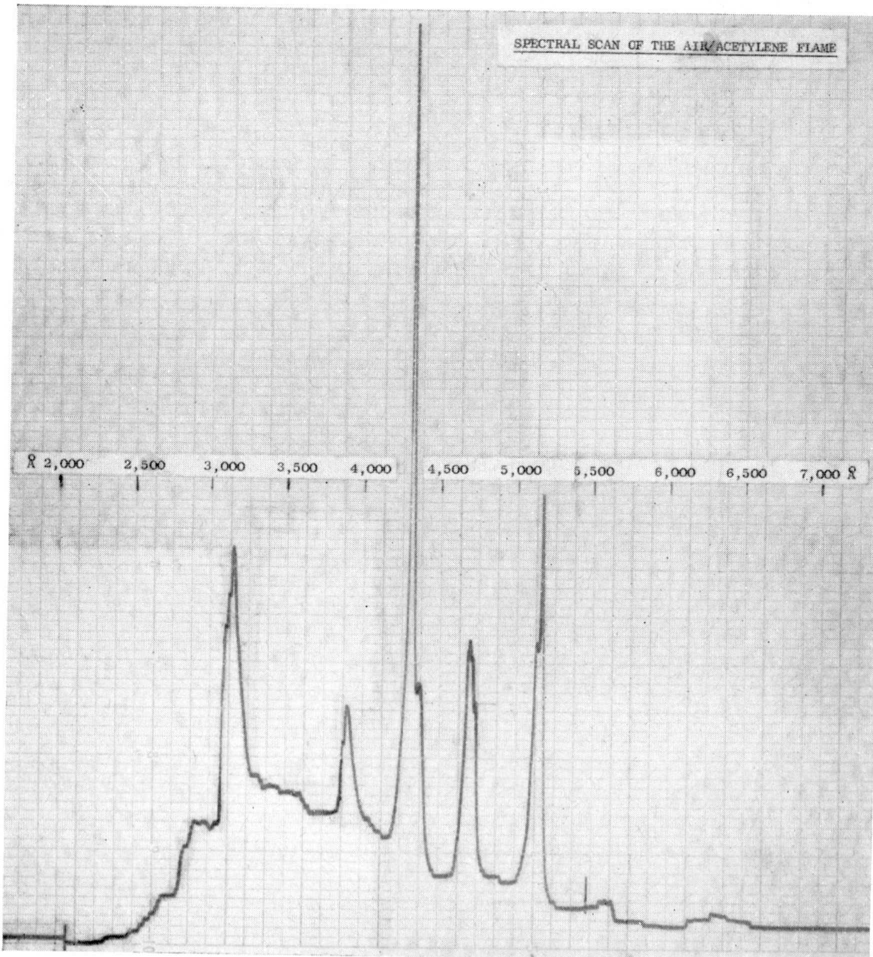

Fig. 2.1 Spectral scan of the air—acetylene flame

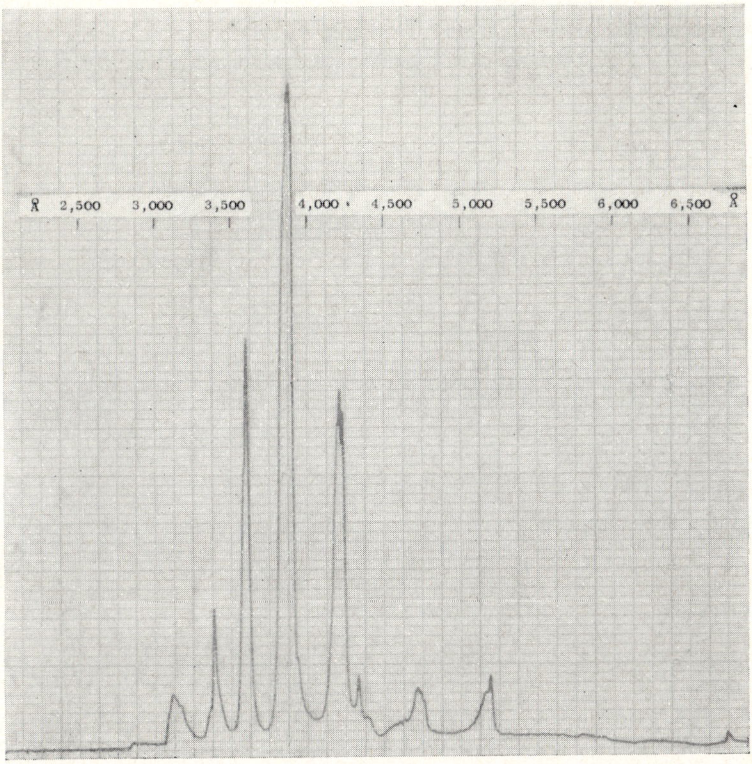

Fig. 2.2 Spectral scan of the nitrous oxide—acetylene flame

flame, and using a hollow cathode lamp of the original type, the background noise at the most sensitive lines at 3579 and 3594 Å is excessively high, so that it becomes essential to use the less absorbing but more stable line at 4254 Å. The improved high spectral output lamps for chromium allow the most sensitive lines to be used for determinations with the nitrous oxide–aceytlene system.

For antimony the 2312 Å line is slightly less absorbing than the 2068 Å line, but gives rise to a more stable response which is more suitable when scale expansion is to be used for low-level determinations.

It may happen that the linearity (i.e. sensitivity) of the most absorbing line drops off very severely as the concentration of the absorbing species increases. An example of this phenomenon is provided by nickel (see Fig. 2.3) for which both the more sensitive 2320 Å and 3415 Å lines are less useful for determinations in solutions at concentrations above 150 μg/ml than the less sensitive line at 3051 Å.

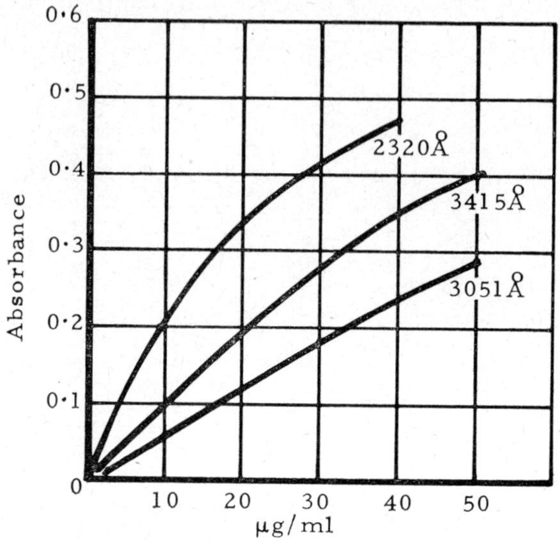

Fig. 2.3 Response curves for nickel at different wavelengths

Solution Conditions

The constitution of standard and test solutions, the third parameter within the analyst's direct control, can also be adjusted to enhance sensitivity. The addition of interference suppressants such as lanthanum chloride or E.D.T.A. disodium salt (again to overcome the depressive effect of phosphate upon calcium in the air–acetylene flame) is a well-established procedure. It is of interest to note that the actions of these two reagents, either of which produce the same result, are different. Lanthanum chloride reacts preferentially to form lanthanum phosphate thereby releasing the calcium atoms for absorption in the flame[4]. E.D.T.A. disodium salt, by contrast, exerts a greater affinity than phosphate towards calcium. The calcium–E.D.T.A. compound so formed then breaks up in the flame to yield calcium atoms which are free to absorb light energy of the correct frequency.

The use of organic solvents that are miscible with water, to enhance an element's response, was at one time a fairly common practice. Industrial methylated spirits, isopropyl alcohol, and acetone were some of the most commonly used. Aqueous mixtures of these materials show little tendency to change composition upon evaporation, have good solvent properties, and give rise to an enhancement that is independent of small changes of solvent composition. The effect of such a solvent is entirely physical. It gives rise to a medium that is more readily and efficiently nebulised so that a greater proportion of small droplets, and hence absorbing species, is carried to the flame.

Determinations that are carried out on extractions made into organic solvents, as, for example, when heavy metals (such as lead) are chelated at dilution in an aqueous medium and then concentrated into isobutyl methyl ketone, gain in sensitivity by both the concentration factor, and the greater aspiration rate of the solvent. This phenomenon also occurs in the direct determination of the metallic constituents of oils and petrols.

Sensitivity and Limit of Detection

Sensitivity

It is common practice, in the literature of atomic absorption spectroscopy, to define the capability of an instrument to determine a particular element by means of a figure called the 'sensitivity for 1 per cent absorption'. This is a theoretical number and will vary with the efficiencies of the lamp, atomiser, flame system, monochromator, and photomultiplier. It will also depend on the solvent and gas mixture employed. Its greatest drawback is that the sensitivity for 1 per cent absorption takes no account of noise level, so it is quite valueless as a guide to the least quantity of an element that can be determined.

The 1 per cent absorption figure should be calculated in the following manner:

1. From a calibration curve of an aqueous solution of the species, obtained without the use of scale expansion, the concentration in μg per ml is read that gives rise to a signal equivalent to an optical density of 0·1.
2. Optical density or absorbance is defined as $\log_{10} I_0/I$ where I_0 is the intensity of incident light and I is the intensity of transmitted light. If 1 per cent of the incident light has been absorbed, then the absorbance of the medium is given by: $\log_{10} 100/99 = 0{\cdot}0044$. Assuming that the absorbance is proportional to the concentration of the absorbing species, in the solution being aspirated to the flame, the concentration ($C_{0\cdot1}$) that gives rise to an absorbance of 0·1 is given by

$$C_{0\cdot1} = K \times 0{\cdot}1$$

where K is a constant.

Likewise

$$C_{1\%} = K \times 0{\cdot}0044$$

where $C_{1\%}$ is the concentration that gives rise to 1 per cent absorption (i.e. the sensitivity for 1 per cent absorption).

$$\therefore \qquad C_{1\%} = \frac{C_{0\cdot1} \times 0{\cdot}0044}{0{\cdot}1} = C_{0\cdot1} \times 0{\cdot}044$$

Thus the sensitivity for 1 per cent absorption is the concentration of an aqueous solution that gives a signal equivalent to an optical density of 0·1, multiplied by 0·044.

It is apparent from the above discussion that the only reliable way to decide whether or not a particular estimation can be carried out is to refer to a typical response curve and to take account of the noise level. The curve will indicate the concentration range over which determination is possible and the slope of the curve and noise level will define the differences that can reliably be detected within the range.

LIMIT OF DETECTION

For low-level determinations of metals in solution it is necessary to resort to the use of scale expansion.

In order to assess whether or not the technique is applicable to a particular low-level determination it is necessary to carry out an actual test run over the range required, using the appropriate expansion facility. From the data obtained from such a run the *true detection limit*, which is usually taken as twice the noise level, can be obtained.

Figure 2.4 (a) and (b) show recorder traces for two elements of about the same sensitivity, one of which (Cd) exhibits a much higher noise level than the other (Li).

Commercial instruments have recently become available in which integrated readings are obtained by allowing the total *fluctuating* current from the photomultiplier to charge a capacitor for a set period (say 15 seconds), followed by measurement of the p.d. of the capacitor. The integrated readings so obtained are thus absolutely steady, so that there is no uncertainty in taking a *single* reading.

Obviously the above criterion for the limit of detection will not apply to such readings. For this reason it has been proposed, by the Atomic Absorption Spectroscopy Group of the Society of Analytical Chemistry, that the limit of detection shall be defined as: 'The minimum amount of an element which can be detected with a 95 per cent certainty. This is that quantity of the element that gives a reading equal to twice the standard deviation of a series of at least ten determinations at or near

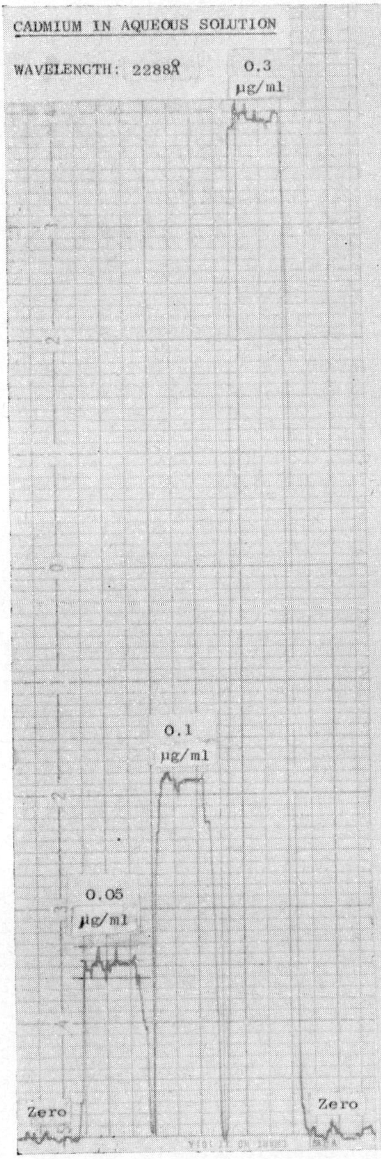

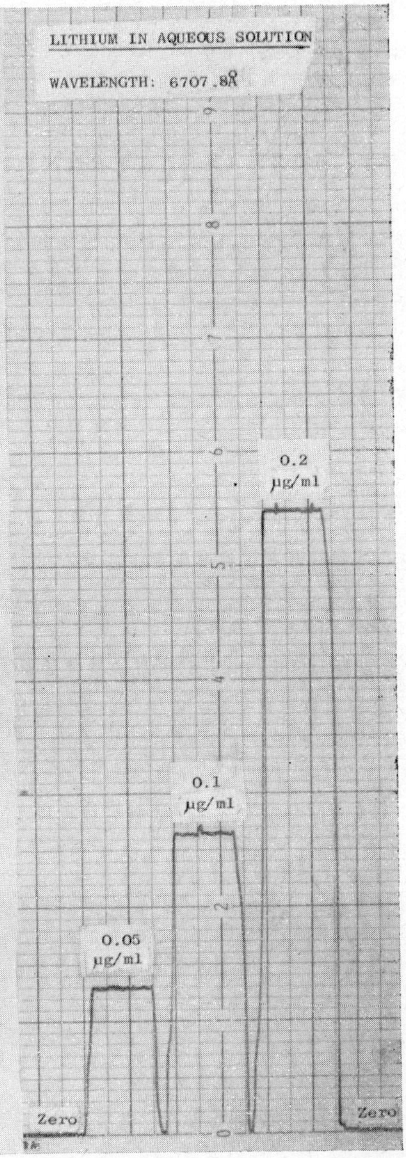

Fig. 2.4 (a) Example of an element showing a fairly high noise level

Fig. 2.4 (b) Example of an element showing a low noise level

the blank level.' This definition can, of course, be equally well applied to readings taken in the 'direct' mode (by measuring the fluctuating output from a photodetector).

Accuracy of Analyses

The instrumental precision with which a determination can be made will vary from element to element according to its sensitivity and the stability of its response. For a given element these functions will themselves vary according to the matrix in which the element is being determined (i.e. the concentrations of foreign species, composition of the solvent, etc.). The final accuracy to which a determination can be reported will be greatly influenced by the dilution factor that was required to prepare a test solution from the sample, which of course is governed by the concentration of the species being determined in the original sample.

The most striking feature of atomic absorption methods is the excellent reproducibility that is obtained upon successively aspirating the same sample solution.

For calcium determinations, work carried out on the basic EEL 140 atomic absorption spectrophotometer, without scale expansion, showed that the standard deviation in instrument reading varied from 0·00179 O.D. units at a concentration of 5 μg/ml to 0·00358 O.D. units at 15 μg/ml leading to an analytical precision of 0·030 μg/ml at the 5 μg/ml level and 0·145 μg/ml at 15 μg/ml.

Chap. II References

1. BILLINGS, G. K., Light scattering in trace-element analysis by atomic absorption. *Atomic Absorption Newsletter* 4 (1965) 357.
2. KIORTYOHANN, S. R., and PICKETT, E. E., Light scattering by particles in atomic absorption spectrometry. *Anal. Chem.* 38 (1966) 1087.
3. WILLIS, J. B., Determination of lead and other heavy metals in urine by atomic absorption spectroscopy. *Anal. Chem.* 34 (1962) 614.
4. YOFE, J., and FINKELSTEIN, R., Elimination of anionic interference in flame photometric determination of calcium in the presence of phosphate and sulphate. *Anal. Chim. Acta* 19 (1958) 166.

III

The Techniques of Measurement

Types of Determinations

THERE is no branch of chemical analysis involving the determination of metals in solution that cannot without advantage use atomic absorption spectroscopy. The technique is already a firmly established procedure in such widely varying fields as clinical chemistry, ceramics, petroleum chemistry, metallurgy, mineralogy, biochemistry, soil analysis, water supplies, and effluents. The general procedures used in performing analyses are conveniently categorised according to the technique required for sample and standard preparation, as shown below:

1. Determinations that can be made against simple standard solutions containing only the element being sought.
2. Determinations that require the addition of a chemical interference suppressant to the sample and standard solutions.
3. Determinations where enhancement by the use of mixed organic–aqueous solvents is exploited.
4. Determinations requiring the use of complex standard solutions prepared so as to match approximately the composition of the test solutions.
5. The use of extractive concentration techniques.
6. Determinations carried out directly and entirely in non-aqueous solvents.

7. Determinations of macro constituents by 'backing off' followed by scale expansion.
8. Indirect determinations, usually exploiting chemical amplification reactions.

DETERMINATIONS MADE AGAINST SIMPLE STANDARD SOLUTIONS CONTAINING ONLY THE ELEMENT (OR ELEMENTS) BEING SOUGHT

Although interferences of one metal upon another, or of excess acid upon a metal's absorption do occasionally occur, atomic absorption spectroscopy is in general remarkably free from such effects. When it is required to develop a procedure for large numbers of similar analyses (particularly when only a few of the elements present in the sample are to be estimated) it is always worthwhile examining the possibility of exploiting this simple approach.

Comparison against simple standards is valid for many analyses and because of its speed is of great value for sighter trials. It is especially likely to be possible when the nitrous oxide–acetylene flame is being used. It is also worth noting that nickel, zinc, iron, copper, and lead are remarkably free of interferences from the elements with which they are likely to be found in association.

Sometimes, indeed, the analyst is prohibited from using complex standards, an example of this situation occurring in the trace determinations of copper, cadmium, and zinc in tin–lead solders. These determinations are complicated by the fact that even the purest lead and tin salts (readily available to the chemist) contain substantial quantities of cadmium and zinc. The preparation of matching standards that contain the two major constituents of the alloy is therefore impracticable. An exhaustive series of recovery trials using that well-known analytical gambit, the method of additions, proved that accurate results could be obtained merely by using standards that contained copper, cadmium, and zinc in concentrations that covered the ranges of these elements in the test solutions.

This elementary procedure finds application in a very wide range of analyses, but is of particular value in the analysis of waters, effluents and soil-extracts, where there is little else in the test solutions other than the elements to be determined and even these are present at low levels.

DETERMINATIONS THAT REQUIRE THE ADDITION OF A CHEMICAL INTERFERENCE SUPPRESSANT TO THE SAMPLE AND STANDARD SOLUTIONS

One of the most firmly established and expanding applications that atomic absorption has found is for the determination of calcium and magnesium in blood serum. The phosphate ion occurs in this fluid at a sufficiently high concentration level to exert a depressive effect upon the calcium absorption when an air–acetylene flame is used. This depression is conveniently overcome by the addition of a heavy excess of lanthanum chloride or E.D.T.A. disodium salt. Comparison must, of course, be made against a blank and standard solutions containing the suppressant.

Strontium chloride may also be used as a releasing agent to overcome phosphate interference, but is less efficient than lanthanum chloride, and the quantity added must be controlled more stringently in order to avoid non-uniform over-enhancement of the calcium response. Figure 3.1 illustrates this. Lanthanum chloride is undoubtedly the most useful of the releasing agents employed to overcome chemical interference. It effectively overcomes the depressive effect of aluminium upon calcium and magnesium. It may also be used to overcome the depression

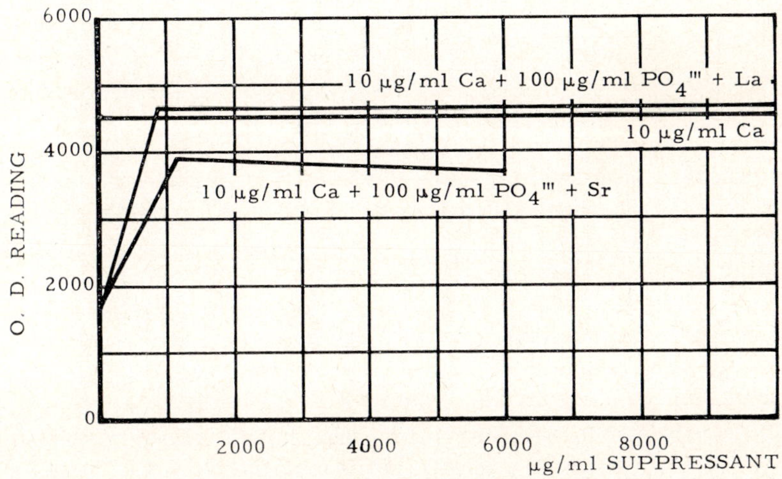

Fig. 3.1 Comparison of lanthanum chloride and strontium chloride as interference suppressants

of iron upon chromium that is so pronounced in the air–acetylene flame. The addition of a heavy excess of an alkali-metal to a solution containing calcium, strontium, or barium, that is to be estimated in the nitrous oxide–acetylene flame may be considered as belonging to this category. Here, though, the interference to be suppressed is a physical one (production of unwanted calcium ions) rather than one of a chemical nature.

Determinations where Enhancement by the Use of Aqueous-Organic Solvents is Exploited

As mentioned previously, it was at one time fairly common practice to utilize mixed organic aqueous solvents, such as isopropyl alcohol and water, with the object of improving the sensitivity for estimations.

The development of scale expansion facilities has not only reduced the need to resort to this procedure, but has also detracted from its usefulness. This comes about because although the use of a mixed organic–aqueous solvent will increase the *sensitivity* for an estimation, it will usually also increase the background noise level. Thus, when scale expansion is utilized, no improvement in detection limit is attained.

For the direct determinations of trace metals in liquors such as whisky comparison must, of course, be made against standards and a blank that also contain alcohol at the same level as the whisky.

Determinations requiring the Use of Complex Standard Solutions, prepared so as to match approximately the Composition of the Test Solutions

Determinations in which measurement is made against complex standards prepared to resemble, in constitution, the solution to be analysed are perhaps the most common carried out by the analyst. Most chemical interferences reach a plateau value, after which further addition of the interfering substance has no effect upon the element to be determined. Examples of the types of analyses tackled by this procedure are multitudinous, but the technique is especially valuable for the analysis of minerals, ores, silicates, ceramics, slags, and alloys.

Apart from the main intention of overcoming chemical interferences the procedure reduces the number of separate standard solutions required. For example in the analysis of a soil extract containing calcium, magnesium, sodium, and potassium, although none of the elements present interferes with the absorption of any of the others, when the air–acetylene flame is used, complex standards containing all of them will reduce the number of flasks containing solutions from, say, 12 to 3 (if three concentrations for each element are prepared to cover the expected ranges in the test solutions).

A convenient example of the use of complex standards to counteract interference effects, and also to rationalise the number of solutions to be prepared, is provided by the determination of the minor constituents of a duralumin alloy. The alloy would perhaps have the following composition:

Cu	4·42 per cent	i.e.	44200 μg/ml
Mg	0·74 per cent	i.e.	7400 μg/ml

Mn	0·73 per cent	i.e.	7300 μg/ml
Zn	0·11 per cent	i.e.	1100 μg/ml
Fe	0·40 per cent	i.e.	4000 μg/ml
Al	92·55 per cent	i.e.	925500 μg/ml
Other trace constituents	1·05 per cent	i.e.	10500 μg/ml

The analyst could be required to estimate the copper, magnesium, manganese, zinc, and iron contents. In order to determine these it would be necessary to prepare test solutions that contain them in the following ranges.

Cu	0·1 to 10	μg/ml
Mg	0·05 to 2·5	μg/ml
Mn	0·1 to 5·0	μg/ml
Zn	0·05 to 4·0	μg/ml
Fe	0·3 to 25	μg/ml

By dissolving 0·5 g of the alloy in hydrocholoric acid and adjusting the volume accurately to 1 litre a master test solution of the following composition would be obtained:

Cu	22·1 μg/ml
Mg	3·7 μg/ml
Mn	3·65 μg/ml
Zn	0·55 μg/ml
Fe	2·0 μg/ml
Al	462·8 μg/ml

Standard solutions would now be prepared that matched this test solution in constitution and covered the anticipated concentration ranges of the elements in it, namely:

	Standard solution I (μg/ml)	Standard solution II (μg/ml)	Standard solution III (μg/ml)
Cu	20	22·5	25
Mg	2·0	3·0	4·0
Mn	2·0	3·0	4·0
Zn	0·2	0·5	1·0
Fe	1·0	2·0	3·0
Al	400	450	500

The minor constituents whose analyses were not required would be omitted from these standards. Aluminium, although its analysis was not required, would be included as the major possible source of interference.

The estimations of all the elements could now be made by aspirating the standard and test solutions to the atomic absorption spectrophotometer, constructing concentration/response curves and reading off the concentrations of the test solutions.

The estimations of iron, manganese, and zinc would most conveniently be made with the air–acetylene flame, using a 10 cm burner. Scale expansion could be used to improve the accuracy for the iron determination.

Burner rotation could be used to reduce the sensitivity of the instrument for the copper determination, or alternatively measurement could be made at the less absorbing 3275 Å line, so as to avoid the necessity for further dilutions. Magnesium would best be determined with the nitrous oxide–acetylene flame, and again burner rotation could conveniently be used.

The Use of Extractive Concentration Techniques[1–4]

Determinations that involve prior extraction of the species to be determined into an organic solvent have been widely employed. They are particularly valuable for the determination of trace metals in water, effluents, and extracts from food and botanical samples. Extraction also provides a useful means of separating metals such as lead and zinc from solutions that contain heavy loadings of foreign materials, the presence of which can cause errors due to 'light scatter' at wavelengths below 2200Å.

The most useful method for extractive concentration of heavy metals is that employing chelation with ammonium pyrrolidine dithiocarbamate (A.P.D.C.) followed by extraction into isobutyl methyl ketone (I.B.M.K.). The A.P.D.C. chelation complexes of the common metals are formed over a wide *p*H range, e.g. Cr and Mo, *p*H 2–6; Sn, *p*H 2–8; Mn^{II}, Fe, Co, Ni, Cu, Zn, Pb and Cd, *p*H 2–14. Most authorities recommend a *p*H of between 2·2 and 2·8 for the extractions.

The clear aqueous solution that contains the metal(s) to be determined, and which may have been previously reduced in volume by evaporation, is transferred to a separating funnel. The *p*H is adjusted by the addition of 0·5 N hydrochloric acid or 0·5 N sodium hydroxide solution to lie in the range 2·2 to 2·8. One ml of a 1 per cent aqueous solution of A.P.D.C. is added, followed by 10 ml of I.B.M.K. The vessel is stoppered and shaken for two minutes, the phases allowed to separate, and the aqueous layer run to waste.

The trace heavy metals are now determined in the organic extract against standards that cover the appropriate range and have also been prepared by chelation and extraction as described above. The absorption obtained upon spraying the extracts to an atomic absorption spectrophotometer is enhanced by both the concentration factor and the greater volatility of the solvent. A recorder trace illustrating the sensitivity for lead extracted into I.B.M.K. is shown in Fig. 3.2.

DETERMINATIONS MADE DIRECTLY AND ENTIRELY IN NON-AQUEOUS SOLVENTS

One of the major advantages that atomic absorption spectroscopy offers to the chemist is that solutions need not be confined to an aqueous medium. This means that digestions or extractions that are so time-consuming, necessitate the running of reagent blanks, and are always open to the risk of handling losses, are avoided. The use of non-aqueous solvents is especially convenient in the field of petroleum analysis. The direct determinations of aluminium and vanadium in crude and fuel oils is readily accomplished, as are those of additives (zinc, calcium, and barium) and even wear metals in lubricating oils.

It is, of course, necessary to dilute oils and greases with a suitable solvent before they can be aspirated to an atomic absorption spectrophotometer. References will be found in the literature that recommend the use of iso-octane, n-heptane, xylene, and other hydrocarbons for this dilution. It is my (R.J.R.'s) experience though, (which is mainly confined to the use of the EEL Models 140 and 240) that such solvents are unsuitable when used alone, since they give rise to such a rich smoky flame that it is found necessary, in order to establish non-luminous conditions, to restrict the acetylene flow to the point where the flame eventually lifts off the burner. Mixtures of white spirit and acetone, white spirit and isopropyl alcohol, and iso-octane and isobutyl methyl ketone overcome this difficulty. Cyclohexanone, by itself, is also suitable and with these diluents, solutions may be prepared that allow determinations to be carried out with both the air–acetylene and nitrous oxide–acetylene flames with safety.

The standards for such determinations must, of course be prepared from suitable oil-soluble compounds, or from previously analysed oils, so as to resemble the test solutions.

The analysis of lead in petrol constitutes a special and interesting case. The two compounds most commonly employed as anti-knock additives are tetraethyl and tetramethyl lead. The direct determination

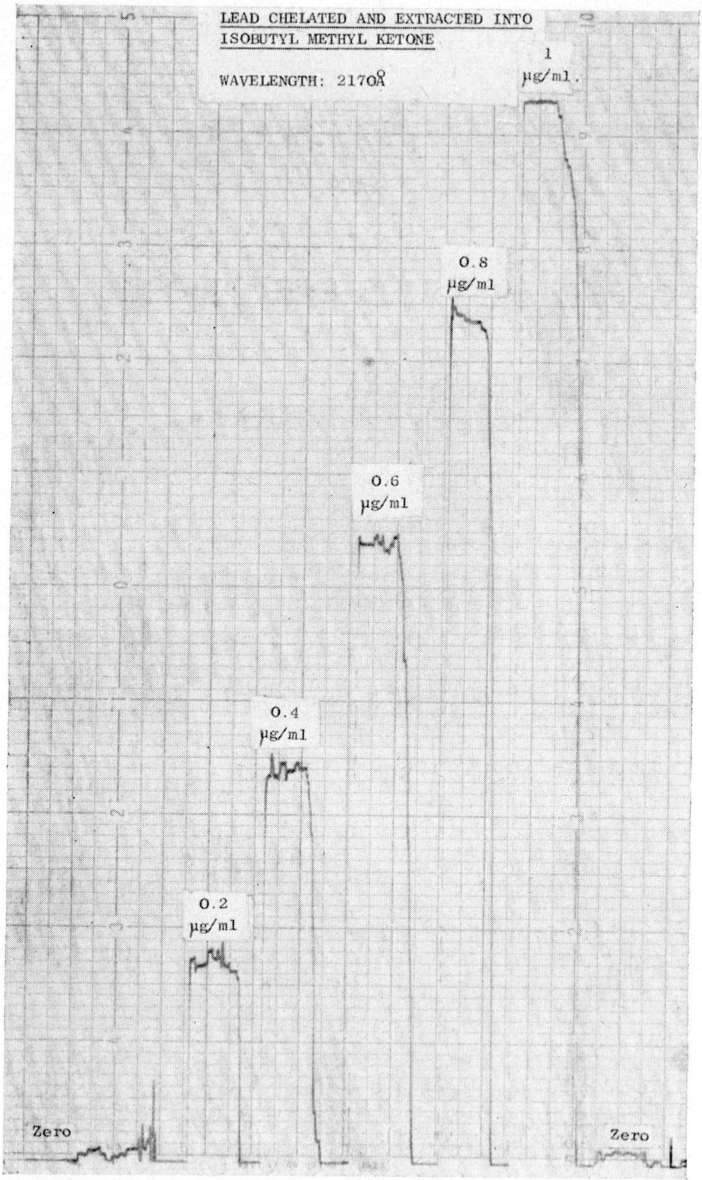

Fig. 3.2 Lead chelated and extracted into isobutyl methyl ketone

is complicated by the fact that the absorption exhibited for the same quantity of lead when present as T.E.L. differs greatly from that shown when present as T.M.L. Thus, in order to perform a direct quantitative analysis of lead in petrol, it is necessary to know which of these compounds, or what mixture of the two, has been included in the sample.

It is standard practice in the petroleum industry to use mixtures of T.E.L. and T.M.L., prepared so that they have the same lead contents as the pure tetraethyl or tetramethyl master compounds. Blenders can take advantage of this fact to check that blends have been properly constituted. A family of curves can be constructed from dilutions of various T.E.L./T.M.L. mixtures whose relative proportions are known. Works batches could then be compared against these curves, and an assessment of their relative T.E.L. and T.M.L. contents made.

If the analyst is faced with a determination of lead in a petrol, and possesses no knowledge as to the nature of the lead additive, then direct determination is not valid. In such cases, destruction of the lead alkyl, followed by extraction of the lead into an aqueous phase, is necessary.

Determination of Metals present as Macro Constituents of a Sample

An example of this type of analysis is provided by the determination of calcium in cement, where that element comprises about 46·5 per cent of the cement.

Preliminary work establishes that the optimum concentration range for the determination of this element is between 4 and 6 μg/ml. If a 2 g sample of cement is used, this will need to be dissolved and effectively diluted by a factor of 100 000 in order to obtain a test solution that contains calcium at the 4 to 6 μg/ml level. In order to achieve the required accuracy it is necessary, therefore, to be able to discern very small differences of Ca in the test solutions. This requirement is achieved by 'backing off' the signal so that a 4 μg/ml standard gives a reading of zero on the meter of the atomic absorption spectrophotometer. Scale expansion is then applied over the range 4 to 6 μg/ml Ca. This procedure will, under the best conditions, permit a standard deviation of 0·02 μg/ml Ca to be attained, giving rise to an accuracy of ±0·2 per cent on the cement.

Other determinations of the same type that have been performed are the determinations of iron and manganese in ferrites that contained about 45 per cent Fe and 14 per cent Mn. The procedure is applicable also to metallurgical analyses.

Indirect determination using Chemical Amplification Procedures[5]

Interest in amplification (or multiplication) reactions has revived recently in several branches of analytical chemistry because they allow trace analyses to be performed to a very high degree of precision. In the field of atomic absorption spectroscopy, West and his school at Imperial College, London, have applied indirect amplification reactions, in which molybdenum complexes are produced, to the determinations of phosphate, silicate, and niobium.

Taking the determination of phosphate as an example, the principle of the method lies in the formation, and extraction into an organic phase, of phosphomolybdic acid $H_3PO_4(MoO_3)_{12}$. The molybdenum is then determined by atomic absorption, and the estimation benefits from the fact that 12 molybdenum atoms are associated with every atom of phosphorus.

The procedure developed by West and his associates for phosphate and silicate allows the two radicles to be determined sequentially from a solution in which they are both present. The test solution is acidified, treated with an excess of molybdate reagent, and the phosphomolybdic acid preferentially extracted away from the excess reagent and silicate into isobutyl acetate. Estimation of the molybdenum content by atomic absorption is made on the organic extract.

The residual aqueous phase, which contains the silicate, is treated with ammonium hydroxide to lower the acidity and the silicomolybdic acid $H_4SiO_4(MoO_3)_{12}$ is extracted into butanol. The organic phase is washed free from excess molybdate reagent and its molybdenum content determined by atomic absorption spectroscopy.

The same group of workers[6] have developed a similar procedure for the determination of niobium, in which molybdoniobophosphoric acid is formed and extracted into butanol. The phosphomolybdic acid, which is also formed, is previously removed from the solution by selective extraction into isobutyl acetate. In this case eleven molybdenum atoms are associated with every atom of niobium. The procedure is of particular importance, as tantalum does not form a similar complex.

The Method of Additions

The method of additions is a well-known procedure in analytical chemistry and can be useful for checking the accuracy of a determination carried out in the presence of foreign substances, where interference effects upon the element being determined are unknown. For it to be applicable to atomic absorption spectroscopy the meter reading must bear a linear relationship to the concentration of the element in the solution. In addition to the necessity for a linear response/concentration relationship, the method suffers from the disadvantage that one or two additional test solutions must be prepared for each sample.

The procedure will not correct for errors due to light-scattering interference. As stated above the method can be useful for checking the validity of certain determinations and for this reason alone it is worthy of brief mention. The procedure consists of dividing the sample into, say, three aliquots. To two of the aliquots known additions are made of the metal to be determined. The third aliquot is retained untreated. The solutions are then aspirated to the atomic absorption spectrophotometer, the meter readings noted and plotted against the concentration increases of the metal, made to the original sample. The curve is extrapolated and the point of intersection with the *x*-axis is taken to indicate the concentration of the sample. The principle of the method is illustrated in Fig. 3.3.

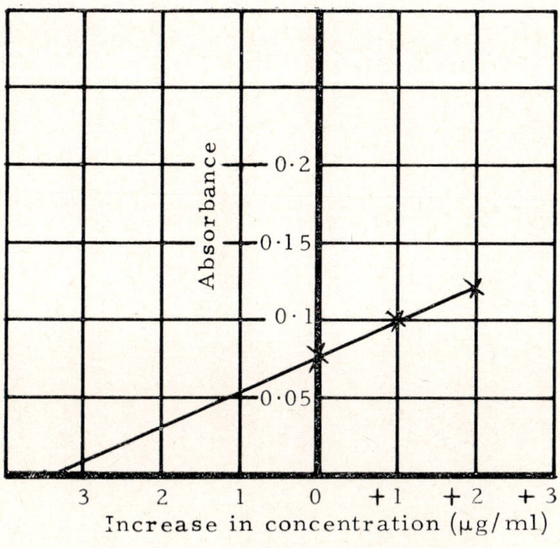

Fig. 3.3. Method of additions

Chap. III References

1. ALLAN, J. E., The use of organic solvents in atomic absorption spectrophotometry. *Spectrochim. Acta* 17 (1961) 467.

2. MALISSA, H., and SCHOFFMANN, E., The use of substituted dithiocarbamates in Microanalysis III. *Mikrochim. Acta* 1 (1955) 187.

3. DEAN, J. A., and LADY, J. H., Application of organic solvent extraction to flame spectrophotometry. *Anal. Chem.* 27 (1955) 1533.

4. LAKANEN, E., Separation and concentration of trace metals by means of pyrrolidine dithiocarbamic acid. *Atomic Abs. Newsletter* 5 (1966) 17.

5. KIRKBRIGHT, G. F., SMITH, A. M., and WEST, T. S., The indirect sequential determination of phosphorus and silicon by atomic absorption spectrophotometry. *Analyst* 92 (1967) 411.

6. KIRKBRIGHT, G. F., SMITH, A. M., and WEST, T. S., An indirect amplification procedure for the determination of niobium by atomic absorption spectroscopy. *Analyst* 93 (1968) 292.

IV

Characteristics of the Elements

In this chapter details of procedure are given for all the elements that can be determined by atomic absorption spectroscopy. Approximate and abbreviated data are tabulated for convenient reference in Tables IV.1 and IV.2, and fuller and more precise information follows.

Aluminium

WAVELENGTHS

The most absorbing resonance lines in the aluminium spectrum are at 3093 and 3962 Å. They are of about equal sensitivity, but a higher intensity of light is obtained at the 3962 Å line.

FLAME SYSTEM

It is necessary to use the nitrous oxide–acetylene flame.

SENSITIVITY

A sensitivity of 1·0 μg/ml for 1 per cent absorption can be expected with standard instruments.

The limit of detection at a scale expansion of ×10 is about 0·2 μg/ml in aqueous solution.

In aqueous solution the optimum range for determination is from about 2 to 500 μg/ml and, without scale expansion, differences of about 2 μg/ml can be estimated. In a medium of 1+1 isopropyl alcohol and water, aluminium can be determined over the range 1 to 250 μg/ml and differences of about 1 mμ/ml can be estimated.

Interferences and Special Characteristics

Solutions containing equal quantities of aluminium as the chloride or nitrate give rise to identical responses; but a solution of potash-alum

Table IV-1. Common Elements in Aqueous Solution that can be determined by AAS

Element	Wave-length Å	Flame system	Sensitivity for 1% absorption (μg/ml)	Limit of detection at × 10 scale expansion (μg/ml)	Useful range for determination without scale expansion (μg/ml)
Aluminium	3093	N_2O–acetylene	1·0	0·2	2–500
Aluminium	3962	N_2O–acetylene	1·0	0·2	2–500
Antimony	2068	Air–acetylene	1·0	0·5	2–150
Antimony	2176	Air–acetylene	0·7	0·2	1·5–100
Antimony	2312	Air–acetylene	1·4	0·5	3–200
Arsenic	1937	Argon–hydrogen	1·5	0·5	2·5–100
Barium	4554	N_2O–acetylene	0·8	0·08	1·5–250
Barium (+1000 μg/ml K)	5536	N_2O–acetylene	0·8	0·08	1·5–250
Beryllium	2349	N_2O–acetylene	0·04	0·005	0·1–10
Bismuth	2231	Air–acetylene	0·7	0·1	1·5–75
Boron	2497	N_2O–acetylene	24	10	50–3000
Cadmium	2288	Air–acetylene	0·04	0·01	0·1–5
Caesium	8521	Air–acetylene	0·5	0·04	1–100
Calcium	4227	Air–acetylene	0·1	0·01	0·2–15
Calcium (+1000 μg/ml K)	4227	N_2O–acetylene	0·02	0·008	0·05–4
Chromium	3579	Air–acetylene	0·08	0·01	0·25–20
Chromium	3579	N_2O–acetylene	0·15	0·07	0·4–30
Cobalt	2407	Air–acetylene	0·2	0·03	0·4–40
Copper	3248	Air–acetylene	0·05	0·005	0·1–15
Gallium	2874	Air–acetylene	2·0	0·5	5·0–250
Germanium	2652	N_2O–acetylene	2·0	1·0	4·0–200
Gold	2428	Air–acetylene	0·3	0·05	0·5–20
Indium	3039	Air–acetylene	1·0	0·5	2·0–200
Iridium	2640	Air–acetylene	20	3·0	40–8000
Iron	2483	Air–acetylene	0·1	0·03	0·2–40
Lead	2170	Air–acetylene	0·3	0·03	0·4–40
Lead	2833	Air–acetylene	0·5	0·08	1·0–100
Lithium	6708	Air–acetylene	0·04	0·004	0·1–5
Magnesium	2852	Air–acetylene	0·005	0·001	0·02–2
Manganese	2800 (triplet)	Air–acetylene	0·05	0·005	0·2–10
Mercury[I]	2537	Air–acetylene	1·0	0·2	2·0–100
Molybdenum	3133	Air–acetylene	0·6	0·2	1·5–300
Molybdenum	3133	N_2O–acetylene	0·6	0·2	1·5–300
Nickel	2320	Air–acetylene	0·1	0·02	0·2–25
Nickel	3415	Air–acetylene	0·5	0·05	1·0–50
Palladium	2476	Air–acetylene	0·5	0·1	1·0–125
Platinum	2659	Air–acetylene	3·0	0·5	6·0–500

Element	Wave-length Å	Flame system	Sensitivity for 1% absorption (μg/ml)	Limit of detection at ×10 scale expansion (μg/ml)	Useful range for determination without sca expansion (μg/ml)
Potassium	7665	Air–acetylene	0·02	0·002	0·04–5
Rhodium	3435	Air–acetylene	1·0	0·1	2·0–100
Rubidium	7800	Air–acetylene	0·08	0·01	0·2–10
Selenium	1960	Argon–hydrogen	0·4	0·15	1–50
Silicon	2516	N_2O–acetylene	2·0	0·3	4·0–800
Silver	3281	Air–acetylene	0·05	0·008	0·1–10
Sodium	5890	Air–acetylene	0·02	0·002	0·04–5
Strontium	4607	Air–acetylene	0·1	0·02	0·2–20
Tellurium	2143	Air–acetylene	0·5	0·1	1–50
Tin	2246	Air–hydrogen	0·4	0·05	1·0–50
Tin	2246	N_2O–acetylene	2·4	0·3	5·0–300
Tin	2863	Air–hydrogen	0·8	0·1	2·0–100
Tin	2863	N_2O–acetylene	4·5	0·6	10·0–600
Titanium	3643	N_2O–acetylene	3·0	0·4	6·0–500
Tungsten	2551	N_2O–acetylene	10	5	20–4000
Vanadium	3184 (triplet)	N_2O–acetylene	2·0	0·2	4·0–750
Zinc (pure zinc cathode)	2139	Air–acetylene	0·02	0·004	0·04–2·0
Zinc (brass cathode)	2139	Air–acetylene	0·05	0·010	0·1–5·0

TABLE IV-2. LESS COMMON ELEMENTS THAT HAVE BEEN DETERMINED BY AAS

Element	Wave-length Å	Flame system	Sensitivity for 1% absorption
Dysprosium	4212	Nitrous oxide–acetylene	1·0
Erbium	4008	Nitrous oxide–acetylene	1·5
Europium	4594	Nirtous oxide–acetylene	0·15
Gadolinium	3684	Nitrous oxide–acetylene	40
Hafnium	3073	Nitrous oxide–acetylene	15
Holmium	4104	Nitrous oxide–acetylene	1·0
Lanthanum	5501	Nitrous oxide–acetylene	100
Lutecium	3360	Nitrous oxide–acetylene	15
Neodymium	4925	Nitrous oxide–acetylene	20
Niobium	3349	Nitrous oxide–acetylene	25
Osmium	2909	Nitrous oxide–acetylene	2·0
Praseodymium	4951	Nitrous oxide–acetylene	75
Rhenium	3461	Nitrous oxide–acetylene	10
Ruthenium	3499	Air–acetylene	1·5
Scandium	3912	Nitrous oxide–acetylene	1·0
Samarium	4297	Nitrous oxide–acetylene	15
Tantalum	2715	Nitrous oxide–acetylene	10
Terbium	4327	Nitrous oxide–acetylene	10
Thallium	2768	Air–acetylene	0·5
Thulium	3718	Nitrous oxide–acetylene	0·5
Uranium	3585	Nitrous oxide–acetylene	120
Yttrium	4102	Nitrous oxide–acetylene	5·0
Ytterbium	3938	Nitrous oxide–acetylene	0·15
Zirconium	3601	Nitrous oxide–acetylene	15

containing the same aluminium loading produces a distinctly higher absorption.

Excess hydrochloric acid when present at the 1·0 N level does not affect the absorption figure, but a slight depression is produced when the acid concentration is raised to 5 N.

The presence of heavy excesses of Ca, Mg, Cu, Fe, Zn, Si and PO_4^{3-} have negligible influence upon the absorption of aluminium. Distinct enhancement occurs in the presence of heavy excesses of K, Na and La, possibly due to suppression of ionisation. For this reason it is good practice when determining aluminium by atomic absorption, to add about 0·1 per cent sodium or potassium to test and standard solutions.

Antimony

Wavelengths

The resonance line at 2068 Å is quoted in the earlier atomic absorption literature as giving rise to the most sensitive conditions. Determinations made at 2312 Å are slightly less sensitive than those made at 2068 Å, but the signal is much more stable. More recent work has shown that the line at 2176 Å is in fact the most absorbing wavelength, but in order to use it effectively the instrument must be capable of resolving it from the non-absorbing line at 2179 Å.

Flame Systems

Either the air–acetylene or air–propane flames can be used with equal advantage.

Sensitivity

For most standard atomic absorption instruments, sensitivities for 1 per cent absorption at the three lines should be expected to be of the following order: 2068, 1·0 μg/ml; 2176, 0·7 μg/ml; 2312, 1·4 μg/ml.

The limit of detection at a scale expansion of $\times 10$ is about 0·5 μg/ml at 2068 and 2312 Å and 0·2 μg/ml at 2176 Å.

Without scale expansion, at 2068 Å the element can usefully be determined in aqueous solution over the range 2 to 150 μg/ml and differences of about 2 μg/ml can reliably be estimated. At 2312 Å the optimum range over which determinations can be carried out is 3 to 200 μg/ml and increments of 3 μg/ml can be discerned. At 2176 Å determinations can be made over the range 1·5 to 100 μg/ml without scale expansion.

Interferences and Special Characteristics

With the air–acetylene flame system, solutions of the chloride, or potassium antimony tartrate, containing equal concentrations of antimony give rise to similar responses.

The presence of the following elements exerts no effect upon the absorption for antimony: Al, Ba, Bi, Ca, Mg, Cr, Pb, Mn, Ni, K, Na, Sn, Sr. Zinc, when present as the nitrate, has no effect upon the antimony absorption, but the presence of zinc chloride exerts a distinct depression. The presence of heavy excesses of Cd, Cu, and Fe cause slight depression of the antimony absorption. Excess hydrochloric acid gives rise to a slight enhancement, but this levels out as the acid concentration increases.

With the air–propane flame system, interference effects occur only in the presence of high concentrations of Bi, which gives rise to a pronounced enhancement, and zinc (as zinc chloride) which causes slight depression.

To ensure the reliable determination of antimony by atomic absorption spectroscopy the analyst must approximately match the acidity and metal composition of the standard solutions to those of the test samples.

Arsenic

Wavelengths

The resonance line at 1937 Å is the one most likely to be of practical value to the analyst. There is a slightly less sensitive line at 1972 Å.

Flame Systems

All flame systems themselves absorb strongly at wavelengths below 2000 Å. Flames using hydrogen as a fuel are more transparent than those using acetylene, so are likely to be advantageous for the determination of arsenic. Recent studies with the diffuse argon–hydrogen system have produced a sensitivity for 1 per cent absorption of 1·5 μg/ml and a limit of detection of 0·5 μg/ml.

Interferences and Special Characteristics

Hollow-cathode lamps capable of providing a sufficiently intense light output for estimations to be made with standard commercial equipment are now available. Unfortunately, though, the determination of this element is frequently required at very low levels in solutions containing heavy loadings of foreign ions, e.g. sodium. In these conditions very severe 'light scatter' occurs, rendering simple direct estimations of arsenic at low levels by atomic absorption impossible for many applications.

This difficulty is not likely to be alleviated by the use of microwave-excited electrodeless discharge tubes, or by the replacement of atomic absorption by atomic fluorescence as the measuring technique.

Barium

Wavelength

The atomic resonance line at 5536 Å and the ionic line 4554 Å are the most sensitive.

Flame System

For realistic determinations it is necessary to use the nitrous oxide–acetylene flame.

Sensitivity

The sensitivity for 1 per cent absorption for standard instruments is about 0·8 μg/ml at either of the above wavelengths.

The limit of detection at a scale expansion of $\times$ 10 is about 0·08 μg/ml.

The optimum range for determination in aqueous solutions is from about 1·5 to 250 μg/ml over which, without the use of scale expansion, differences of 1·5 μg/ml can be distinguished.

Interferences and Special Characteristics

Barium exhibits a much lower sensitivity to atomic absorption than either calcium or strontium and the use of the nitrous oxide–acetylene flame is essential for realistic determinations.

The element possesses two useful absorbing lines, one a resonance line, 5536 Å, and the other an ionic line, 4554 Å. If the 5536 Å line is used it is necessary in order to attain optimum absorption, to dose the sample and standard solutions with about 1000 μg/ml K. Indeed, if barium is to be determined in the presence of an easily ionisable element, it is essential to use this wavelength. At the 4554 Å ionic line the presence of an alkali metal depresses the barium absorption. In the *absence* of an alkali metal, though, a very stable and slightly more sensitive response is obtained than at the higher wavelength.

Equal concentrations of barium present in solution as either the chloride or nitrate give rise to identical absorptions.

The presence of heavy loadings of Ca, Sr, Mg, Fe, and Zn do not affect the barium response.

Beryllium

Wavelength

2349 Å.

Flame System

For practical purposes the only system of value is nitrous oxide–acetylene.

Sensitivity

With standard atomic absorption equipment a sensitivity for 1 per cent absorption of about 0·04 μg/ml can be attained. At a scale expansion of ×10 the limit of detection with standard equipment is about 0·005 μg/ml.

Without scale expansion the element can be determined in aqueous solution over a range from 0·1 to 10 μg/ml and differences of 0·1 μg/ml can be reliably estimated.

Interferences and Special Characteristics

The absorption of beryllium is unaffected by the presence of heavy loadings of Ca, Mg, and Cu.

Aluminium in heavy excess slightly depresses the beryllium response.

Bismuth

Wavelengths

The line at 2231Å provides the highest sensitivity for solutions containing up to 75 μg/ml Bi. The 3068 Å line is useful for the determination of bismuth in solutions containing up to about 200 μg/ml.

Flame Systems

Either the air–acetylene or air–propane flame may be used. At 3068 Å using the air–acetylene flame, strong hydroxyl emission occurs. This disadvantage is lessened if the air–propane system is used.

Sensitivity

The sensitivity for 1 per cent absorption at 2231 Å obtained on standard instruments is about 0·7 μg/ml. At 3068 Å it is about 2μg/ml. At 2231 Å the limit of detection at ×10 scale expansion is about 0·1 μg/ml. At this wavelength, without scale expansion the element in aqueous solution can be determined over a range from 1·5 to 75 μg/ml and differences of about 1·5 μg/ml distinguished.

At 3068 Å determinations can be made without scale expansion over a range from 4 to 200 μg/ml and increments of about 4 μg/ml reliably estimated.

INTERFERENCES AND SPECIAL CHARACTERISTICS

Of the common elements likely to be found in association with bismuth the following have no effect, even when present in heavy excess, upon that element's absorption in either the air–acetylene or air–propane flames: Sb, Ba, Cd, Cu, Fe, Pb, Sn, Zn.

Excess hydrochloric acid gives rise to a small but noticeable increase in absorption, and it is necessary to match the acidity of standard solutions to that of test solutions.

Boron

WAVELENGTH

The line at 2497·7 Å is stated to be more sensitive than that at 2496·8 Å. With standard commercial instruments, though, the two lines cannot be resolved, so that the mean sensitivity has to be accepted for determinations.

FLAME SYSTEM

Nitrous oxide–acetylene, fuel rich.

SENSITIVITY

The sensitivity for 1 per cent absorption attainable with standard atomic absorption equipment is at best about 24 μg/ml.

Because of the low sensitivity of the element towards atomic absorption it will be necessary to utilize scale expansion for most determinations.

A realistic value for the limit of detection in aqueous solution at $\times 10$ scale expansion is between 10 and 20 μg/ml.

Cadmium

WAVELENGTH

2288 Å.

FLAME SYSTEM

Air–acetylene. Air–propane and air–hydrogen are also suitable, but have no advantages over air–acetylene.

Sensitivity

The sensitivity for 1 per cent absorption is about 0·04 μg/ml on standard commercial instruments. The limit of detection at ×10 scale expansion is 0·01 μg/ml. In aqueous solution without the use of scale expansion the optimum range for determination is from about 0·1 to 5 μg/ml over which differences of 0·1 μg/ml can be distinguished.

Interferences and Special Characteristics

Cadmium is one of the most sensitive of all metals to determination by atomic absorption spectroscopy.

Solutions of the chloride, sulphate, or nitrate, containing the same concentrations of cadmium all give rise to identical responses.

The following elements, even when present in heavy excess, do not affect the absorption of cadmium: Sb, Cu, Fe, Pb, Zn, Sn, Ag. The presence of excess hydrochloric acid slightly enhances the cadmium response, but this effect levels out and the same enhancement is obtained in solutions that are 1 N and 5 N.

Caesium

Hollow-cathode lamps are not at the present time very satisfactory for caesium, but vapour discharge lamps do provide a convenient radiation source for estimation of the element.

Wavelength

8521 Å. In order to make measurements at this wavelength the instrument must be fitted with a suitable red-end photomultiplier.

Flame System

Either the air–acetylene or air–hydrogen flame systems can be used. Very pronounced ionization depression occurs with the air–acetylene system, and it is essential to dose standard and test solutions with a heavy excess of an easily ionized species (2000 μg/ml K) if this flame is used. Ionization depression does not occur at the temperature of the air–hydrogen flame.

Sensitivity

Under the most favourable conditions at ×1 scale expansion the element can be estimated over the range 1 to 100 μg/ml, and differences of about 1 μg/ml distinguished. The sensitivity for 1 per cent absorption is 0·5 μg/ml.

At ×10 scale expansion the limit of detection is 0·04 μg/ml.

INTERFERENCES AND SPECIAL CHARACTERISTICS

Of the elements commonly found in association with caesium, potassium, sodium, lithium and calcium do not affect the absorption in the air-hydrogen flame. Provided the test and standard solutions are adjusted to contain a heavy excess of an alkali metal the same elements do not interfere with the absorption when the air–acetylene system is used.

Calcium

WAVELENGTH

4227 Å.

FLAME SYSTEMS

With the air–acetylene flame a slightly non-linear response is obtained. Without scale expansion, determinations may be carried out over the range 0·2 to 15 μg/ml, and differences of about 0·2 μg/ml can be estimated.

With the nitrous oxide–acetylene flame, aqueous solutions of calcium give lower responses than in the air–acetylene flame, due to ionization interference. This interference is overcome by loading the solution with potassium chloride. Under these conditions calcium can be determined over the range 0·05 to 4 μg/ml.

SENSITIVITY

Using the air–acetylene flame system the sensitivity for 1 per cent absorption for standard commercial instruments is about 0·1 μg/ml. With the nitrous oxide–acetylene flame it is 0·02 μg/ml.

The limit of detection at a scale expansion of $\times$ 10 is about 0·01 μg/ml for air–acetylene, and 0·008 in the nitrous oxide–acetylene flame.

INTERFERENCES AND SPECIAL CHARACTERISTICS

Air–acetylene Flame System. With this flame the same concentration of calcium present as the nitrate will exhibit a lower response than if present as the chloride. Of the foreign ions commonly found in association with calcium, Ba, La, Sr, and E.D.T.A. disodium salt enhance the calcium response, but this enhancement levels out at high concentrations of the foreign ion.

Heavy excesses of Fe, Mg, Na and K do not affect the absorption of calcium in the air–acetylene flame.

Depression of the calcium response occurs in the presence of Al, Mn and PO_4^{3-}. This interference is overcome (in the cases of Mn and PO_4^{3-}

by the addition of lanthanum chloride or E.D.T.A. disodium salt to the solution. In the fuel-rich air–acetylene flame the extent to which phosphate depression occurs depends on the height at which the light path traverses the flame and at a point high in the flame the interference entirely disappears. This fact is, however of little more than academic interest, since it is much more convincing, reliable, and practical to overcome phosphate interference by the addition of a suppressant such as lanthanum chloride. Bicarbonate alkalinity (in natural waters) will depress the calcium response. This effect, though, is easily overcome by the addition of lanthanum chloride, or by adjusting the *p*H of the solution to lie in the range 1·8 to 3·8. In solutions containing hydrochloric acid, at concentrations above 1N, the calcium response is seriously depressed.

Nitrous Oxide–Acetylene Flame System. At the higher temperature of the nitrous oxide–acetylene flame some of the calcium atoms are further dissociated into calcium ions, which do not absorb the characteristic radiation of the ground-state atom. A depression of the response thereby occurs.

This ionization interference can be overcome by heavily dosing the solution with an easily ionized species such as Na or (preferably) K. When carrying out calcium determinations with the nitrous oxide–acetylene flame it is always good practice to dose the test and standard solutions with (say) 1000 μg/ml K, both to improve the response and also to swamp the effect of any minor differences in composition of the test and standard solutions.

Of the common elements likely to be found in association with calcium those which will enhance the response are Ba, Sr, Na, K, and Mg, due to their ability to supply electrons to the flame system and thereby inhibit the tendency for calcium to ionize.

Equal concentrations of calcium as either the nitrate or chloride exhibit the same responses, while the presence of heavy loadings of PO_4^{3-}, Fe, La, and Mn have no influence on the calcium absorption. An excess of hydrochloric acid has negligible effect upon the absorption. A heavy excess of Al depresses the calcium response, but this depression is overcome by the addition of lanthanum chloride.

REFERENCES

A very large volume of literature has appeared on the determination of calcium by atomic absorption spectroscopy, and it is felt that to give an exhaustive list of references would be more confusing than helpful.

As a matter of courtesy, though, it is right that recognition should be given to two workers whose activities did so much to establish atomic absorption as a tool in the analytical laboratory. WILLIS pioneered the application of the technique to calcium determination in clinical chemistry and DAVID worked in the area of agricultural analysis[7,8,9].

Chromium

WAVELENGTHS

Improvements in lamp technology have drastically enhanced the capabilities of simple commercial atomic absorption spectrophotometers for the estimation of chromium. The two most absorbing lines in the chromium spectrum are located at 3579 and 3594 Å.

With the older argon filled lamps there is a gas line very close to the most sensitive resonance line at 3579 Å and using an instrument with a low reslution monochromator, a grossly bent response curve is obtained at this wavelength. Indeed in practice it is found that with such an instrument the 3594 Å line is apparently more sensitive and gives superior results.

With the newer, neon filled lamps which also have superior outputs, an instrument having a fairly low resolution monochromator possesses a capability equal to one having high resolution optics. With this lamp a low resolution instrument can take advantage of the more absorbing nature of the 3579 Å line.

Furthermore because the argon filled lamps, previously supplied, required the use of fairly wide monochromator slit settings it was impossible to make reliable measurements at either 3579 or 3594 Å with the nitrous oxide–acetylene flame, due to severe flame 'noise' at these wavelengths. Determinations with nitrous oxide–acetylene were therefore restricted to the less sensitive 4254 Å line.

The fact that the spectrum from the newer neon filled lamps is cleaner, and of higher intensity permits the use of very narrow monochromator slit widths, so that measurements can be made with the nitrous oxide–acetylene flame at 3579 Å as well as at the less sensitive 4254 Å line.

Spectral traces from the two types of lamps are shown in Fig. 4.1.

FLAME SYSTEMS

Either the fuel rich air–acetylene or nitrous oxide–acetylene flame can be used.

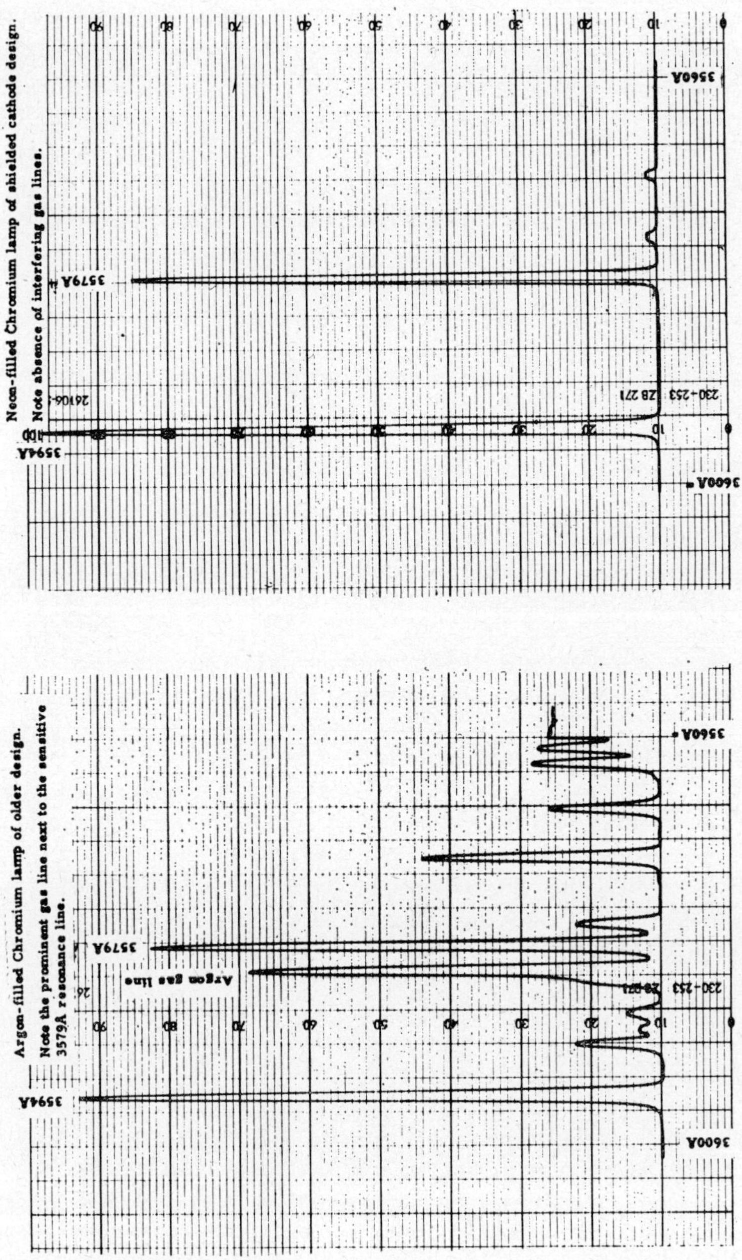

Fig. 4.1 Spectral traces of the older, argon filled, chromium lamps (left) and the newer, neon filled lamps (right)

SENSITIVITY

The essential data for the two flame systems obtained at the 3579 Å line with a neon-filled lamp are tabulated below.

Flame system	Sensitivity for 1% absorption (μg/ml)	Limit of detection at ×10 scale expansion (μg/ml)	Useful range for determination without scale expansion (μg/ml)
Air–Acetylene	0·08	0·01	0·25–20
N_2O–Acetylene	0·15	0·07	0·4–30

INTERFERENCES AND SPECIAL CHARACTERISTICS

Chromium is reasonably sensitive to atomic absorption spectroscopy, but the complexity of its spectrum, and the fact that the presence of iron and vanadium can both cause serious chemical interferences, necessitate its determination to be carried out with particular care.

Air–Acetylene Flame System. The most notable characteristic of the behaviour of chromium in the air–acetylene flame is the chemical interference of iron. This troublesome interference is overcome by the addition of lanthanum chloride or a heavy excess of ammonium chloride. Both of these reagents cause some enhancement of the chromium response and care must be taken to match standard with test solutions when they are used to suppress the effect of iron.

In the air–acetylene flame the same concentrations of chromium present as the alum or chloride give rise to the same absorption. The response for a similar concentration of chromium as the dichromate is higher.

The presence of vanadium, like that of iron, depresses the absorption of chromium.

The following elements have no effect upon the absorption of chromium: Al, Ca, Co, Cu, Mn, Ni, Na. The presence of excess hydrochloric acid does not influence the absortion but excesses of sulphuric and nitric acid do.

Nitrous oxide–Acetylene Flame System. With the nitrous oxide–acetylene flame system similar concentrations of chromium present as the chloride, alum, or dichromate give rise to the same response.

The presence of the following metals has no influence upon the absorption of chromium: Al, Ca, Co, Cu, Fe, Mn, Mo, Ni, V, Na. Once again the presence of excess hydrochloric acid does not affect the absorption, whereas excess of sulphuric or nitric acids does, but to a lesser extent than with the air–acetylene flame.

Cobalt

Wavelengths

2407 Å affords maximum sensitivity, and is most commonly used. Other less sensitive lines occur at 2425, 2531, 3413, and 3527 Å.

Flame System

The air–acetylene flame is most commonly used. Air–propane and air–hydrogen have also been found suitable, but offer no outstanding advantages. With the air–acetylene flame, cobalt can be determined without scale expansion over a range from 0·4 to 40 μg/ml, and differences of about 0·4 μg/ml can be distinguished.

Sensitivity

The sensitivity for 1 per cent absorption for standard commercial instruments is about 0·2 μg/ml. The limit of detection at a scale expansion of ×10 is about 0·03 μg/ml.

Interferences and Special Characteristics

Figure 4.2 compares typical response curves, obtained for aqueous solutions, with two types of lamp, the older type argon-filled lamp and

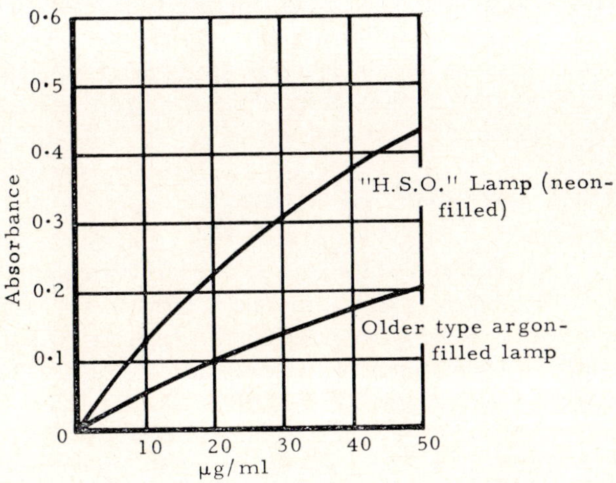

Fig. 4.2 Response curves for cobalt with two types of lamp

the neon-filled high spectral output lamp. These indicate the distinct advantage of employing a high spectral output lamp, which does much to overcome the effect of the complex cobalt spectrum and gives rise to a more sensitive and linear response.

Solutions of the chloride, nitrate or sulphate that contain an equal concentration of cobalt give rise to identical responses.

Of the common metals and radicals likely to be found in association with cobalt, the following do not affect that element's absorption, even when present in heavy excess: Cr, Cu, Fe, Mg, Mn, Mo, Ni, Na, W, V, Si, Ti, and PO_4^{3-}.

Copper

Wavelength

3248 Å is the line most commonly used. That at 3274 Å is less absorbing and can be useful for determinations of solutions containing higher concentrations of copper.

Flame System

The air–acetylene flame is the most commonly used, but the air–propane flame is also suitable.

Sensitivity

The sensitivity for 1 per cent absorption for standard commercial instruments is around 0·05 μg/ml and the limit of detection at a scale expansion of ×10 is about 0·005 μg/ml.

Without scale expansion the element can be determined in aqueous solution at 3248 Å, over a range from about 0·1 to 15 μg/ml and differences of 0·1 μg/ml can be estimated.

Interferences and Special Characteristics

The absorption for copper is remarkably free from interference effects. Solutions of the chloride, sulphate, or nitrate containing the same concentrations of copper exhibit identical responses. The following elements commonly found in association with copper do not influence its absorption, even when present in heavy excess: Al, Ca, Cr, Co, Fe, Pb, Mg, Mn, Ni, Ag, Na, Sn, V, Zn.

With a tubular steel air–acetylene burner, gross non-uniform enhancement occurs in the presence of excess acids. This effect does not occur if a flat-sided laminar flow burner is used.

Gallium

Wavelength

2874 Å is the optimum wavelength for atomic absorption measurements.

Flame System

Air–acetylene.

Sensitivity

The sensitivity for 1 per cent absorption is stated to be 2·0 μg/ml. The useful range for determination in aqueous solution without scale expansion is 5·0–250 μg/ml. At a scale expansion of ×10 the limit of detection in aqueous solution is better than 0·5 μg/ml.

Interferences and Special Characteristics

Heavy loading of Cu, Mg, Zn, Cl′, NO_3', PO_4^{3-}, and SO_4^{2-} do not affect the absorption of gallium, but very heavy loadings of Al produce a slight depression.

Gold

Wavelength

2428 Å.

Flame System

Air–Acetylene.

Sensitivity

The sensitivity for 1 per cent absorption is 0·3 μg/ml.

At a scale expansion of ×10 the limit of detection in aqueous solution is about 0·05 μg/ml.

Without scale expansion the element can be determined in aqueous solution over a range from 0·5 to 20 μg/ml and differences of about 0·5 μg/ml can be reliably estimated.

Interferences and Special Characteristics

As with the noble metals generally, gold is remarkable free from cationic interference effects. The presence of cyanide, though, noticeably depresses the absorption. This depression can be overcome by the addition of lanthanum chloride in heavy excess.

Indium

Wavelengths

The line at 3039 Å exhibits maximum sensitivity. Other less absorbing lines exist at 3256 4105, and 4511 Å.

Flame System

Air–acetylene.

Sensitivity

The sensitivity for 1 per cent absorption is generally quoted to be around 1 μg/ml for standard equipment.

The limit of detection at a scale expansion of $\times$ 10 is about 0·5 μg/ml. Without the use of scale expansion the useful working range is from about 2 to 200 μg/ml, over which differences of 2 μg/ml can be estimated

Interferences and Special Characteristics

Heavy excess of chloride, nitrate, phosphate, sulphate, Al, Cu, Mg, and Zn have negligible influence upon the absorption of indium.

Iridium

Wavelength

The iridium spectrum contains a large number of resonance lines. The one at 2088·8 Å is stated to be the most absorbing but it is of low intensity and a very unfavourable signal-to-noise ratio renders it unsuitable for measurements. Most authorities now recommend the use of the unresolved doublet at 2639·4 and 2639·7 Å (usually shown as 2640 Å) but the lines at 2849·7 and 2924·8 Å are also of value.

Flame System

Air–acetylene.

Sensitivity

The sensitivity for 1 per cent absorption at 2640 Å is about 20 μg/ml. and determinations can usefully be made over a range from 40 to 8000 μg/ml. The limit of detection at $\times$ 10 scale expansion for an aqueous solution is about 3·0 μg/ml.

Interferences and Special Characteristics

By analogy with the other noble metals the absorption for iridium can be expected to be free from interferences by heavy metals.

A solution of an iridite produces a distinctly lower response than that of an iridate.

Iron

Wavelengths

The spectrum of iron possesses a doublet with peaks at 2483 and 2488 Å.

Hollow-cathode lamps for iron are now commonly filled with neon in contrast to the earlier models that contained argon. These newer lamps (containing also an improved electrode configuration) have sufficiently high output to permit the use of very narrow monochromator slit widths. It therefore becomes possible to separate the most absorbing line at 2483 Å from the less absorbing one at 2488 Å. This adjustment requires experience in iron determinations, because the desired peak is not very pronounced. It is best accomplished by approaching from the lower end of the spectrum.

Other less absorbing lines exist at 2167, 2501, 2523, 2719, and 3021 Å, all of which could be of value for the determination of iron at higher concentrations.

Flame System

The air-acetylene flame gives optimum response. The nitrous oxide–acetylene flame can also be used.

Sensitivity

The sensitivity for 1 per cent absorption for standard instruments is about 0·1 μg/ml at the 2483 Å line, using air–acetylene.

The limit of detection at a scale expansion of ×10 is 0·03 μg/ml. In aqueous solution the optimum range for determinations is from 0·2 to 40 μg/ml over which, without the use of scale expansion, differences of 0·2 μg/ml can reliably be estimated.

Interferences and Special Characteristics

Iron absorption is remarkably free from chemical interferences. Of the common substances likely to be found in association with the element Al, Co, Cu, Mg, Ni, K, Na, and PO_4^{3-} cause no alteration of the absorption, even when present in very heavy excess over the iron concentration. Excesses of hydrochloric, nitric and sulphuric acid enhance the absorption, but this effect plateaus out above acid concentrations of 0·5 N.

The presence of silicate depresses the absorption of iron, but this is overcome by the addition of calcium or lanthanum chloride. Aqueous solutions of ferric chloride and ferrous ammonium sulphate containing the same concentrations of iron exhibit the same absorption.

A very noticeable increase in sensitivity is obtained by using mixed aqueous–alcohol solutions.

Lead

Wavelength

The most absorbing line occurs at 2170 Å. The line at 2833 Å is also useful and indeed was the one initially used for lead determinations before the lamps that are now available, possessing improved output characteristics, were produced.

The line at 2614 Å is also of value for determinations upon solutions containing heavy loadings of lead.

Flame System

The air–acetylene flame is the most useful. Determinations can also be made with air–propane but this system offers no advantages in sensitivity over air–acetylene.

Sensitivity

Standard commercial atomic absorption spectrophotometers are generally capable of attaining a sensitivity for 1 per cent absorption of about 0·3 μg/ml at 2170 Å. At 2833 Å the figure is about 0·5 μg/ml.

The limit of detection at a scale expansion of ×10 is 0·03 μg/ml at 2170 Å, and 0·08 μg/ml at 2833 Å.

In aqueous solution the element can be determined at 2170 Å over a range from about 0·4 to 40 μg/ml and, without scale expansion, differences of about 0·4 μg/ml can be estimated. At 2833 Å the optimum range is from 1 to 100 μg/ml over which differences of 1 μg/ml can reliably be distinguished.

Interferences and Special Characteristics

Measurements for lead are remarkably free fom interference effects in both the air–acetylene and air–propane flame systems. Heavy excess of Cu, Fe, Zn, and Sn cause no alteration of the response.

Enhancement attributed to light scattering can be troublesome at the 2170 Å wavelength, and when lead is determined at trace levels in solutions (e.g. urine) containing heavy excess of sodium it is essential that standards should be prepared that approximately match the composition of the test sample as regards sodium content.

Lithium

Wavelength

6708 Å. At this wavelength it is advantageous to use a suitable red-end photomultiplier tube.

Flame System

Either the air–acetylene or air–propane flame is suitable for the determination of lithium.

Sensitivity

The sensitivity for 1 per cent absorption is about 0·04 μg/ml with standard instruments. The limit of detection in aqueous solution at a scale expansion of ×10 is about 0·004 μg/ml.

In aqueous solution without the use of scale expansion the element can be determined over a range from 0·1 to 5 μg/ml and differences of 0·1 μg/ml can be reliably estimated.

Interferences and Special Characteristics

Solutions of the chloride, sulphate, and nitrate that contain the same concentrations of lithium give rise to identical absorptions. The response for lithium is unaffected by the presence of heavy loadings of Al, Ba, Ca, Fe, Mg, K, Na, Zn, and PO_4^{3-}.

Magnesium

Wavelengths

For practical purposes determinations are always carried out at the intense resonance line at 2852 Å.

Flame Systems

The air–acetylene flame is most commonly used, and with this the element can be determined in aqueous solution without scale expansion over a range from 0·02 to 2 μg/ml and differences of about 0·02 μg/ml can be reliably estimated.

Magnesium can also be determined with the nitrous oxide–acetylene flame over a range from about 0·1 to 5 μg/ml in aqueous solution. Increments of about 0·1 μg/ml can be discerned.

Sensitivity

Magnesium is the most sensitive of the elements to determination by atomic absorption.

The sensitivity for 1 per cent absorption in the air–acetylene flame is generally stated to be 0·005 μg/ml and in aqueous solution the limit of detection at a scale expansion of ×10 is less than 0·001 μg/ml. Using the nitrous oxide–acetylene flame the sensitivity for 1 per cent absorption is about 0·05 μg/ml.

Interferences and Special Characteristics

Air–Acetylene Flame. Equal quantities of magnesium, present either as the nitrate or chloride, exhibit the same response, but when present as the sulphate, the sensitivity is slightly diminished. The addition of hydrochloric acid to a solution of magnesium sulphate raises the response to the same level as the response of the chloride solution.

Copper and calcium each exert a small enhancement, while phosphate and aluminium seriously depress the absorption. Phosphate interference can be overcome by the addition of lanthanum chloride.

The presence of Fe, Cr, Mn, Ni, Mo, Pb, Zn, Na, and K have no effect upon the absorption of magnesium in the air–acetylene flame.

Nitrous Oxide–Acetylene Flame. In this flame system solutions of the nitrate, chloride, and sulphate all exhibit identical responses for the same magnesium concentration, and the presence of phosphate does not depress the magnesium absorption.

Calcium, sodium and potassium when present in very heavy excess cause a slight enhancement of the response.

The presence of Al, Cu, Cr, Fe, Pb, Mn, Mo, Ni, and Zn have no effect upon the absorption of magnesium in the nitrous oxide–acetylene flame.

References

A very large number of references to the determination of magnesium in a wide variety of materials will be found in the literature. Allan, Willis, and David contributed largely to the initial development of atomic absorption by their work upon the element.[27,28,29]

Manganese

Wavelengths

There is a triplet in the manganese spectrum with peaks at 2795, 2798 and 2801 Å. Using a normal hollow-cathode lamp it is possible to separate these lines only with an instrument possessing a high performance monochromator. It so happens that all three lines are highly sensitive to atomic absorption so that excellent sensitivity is obtainable with even simple equipment.

A less absorbing line at 4031 Å is of value for determinations of solutions containing manganese up to a level of about 120 μg/ml.

Flame System

Air–acetylene.

Sensitivity

A sensitivity for 1 per cent absorption of 0·05 μg/ml is attainable with standard instruments.

The limit of detection at a scale expansion of $\times$ 10 is about 0·005 μg/ml.

In aqueous solution the optimum range for determinations, without the use of scale expansion, is from about 0·2 to 10 μg/ml, over which differences of 0·2 μg/ml can be estimated.

Interferences and Special Characteristics

Solutions containing the same concentrations of manganese as the chloride, sulphate, nitrate, or permanganate give rise to identical absorptions.

The following cations do not affect the manganese response, even when present in heavy excess: Ca, Cr, Co, Cu, Fe, Pb, Mg, Mo, Ni, K, Na, W, Zn, PO_4^{3-}. In massive excess (e.g. a 1 to 5 per cent solution of the salt) zinc sulphate depresses the response for manganese, whereas zinc chloride at the same level has no effect.

The presence of heavy excesses of Al and Si slightly depresses the manganese absorption, but these interferences can be overcome by adding lanthanum or calcium chloride to the test solution.

It should be noted that iron and nickel salts commonly contain manganese. In order to ascertain reliably the interferences of these two elements upon manganese absorption, therefore, it is necessary to make comparison against a relevant blank.

In the presence of perchloric acid, as for example, when manganese has been extracted by wet digestion from plants or soils, the absorption characteristics depend on the valency of the element. A solution of the permanganate exhibits a depressed absorption, whereas the response for the chloride is unaffected.

For determinations that are carried out in the presence of perchloric acid, therefore, it is essential to reduce the manganese to the divalent state by the action of sodium nitrite or hydroxylamine hydrochloride.

Mercury

Wavelength

The line at 2537 Å is that most commonly used for atomic absorption measurements.

Flame System

Air–acetylene.

Sensitivity

The most notable characteristic is that a solution containing the metal in the mercurous form exhibits a higher absorption than one containing the same concentration of mercury in the mercuric state. This phenomena is explained by the unique characteristic of the mercurous ion to dissociate readily into the mercuric ion and elemental mercury:

$$Hg_2^{2+} \rightarrow Hg^0 + Hg^{2+}$$

Typical response curves obtained with the air–acetylene flame for solutions of mercurous and mercuric salts are shown in Fig. 4.3. It is seen that in the mercuric state the element can be determined without scale expansion over the range 10 to 500 μg/ml and differences of

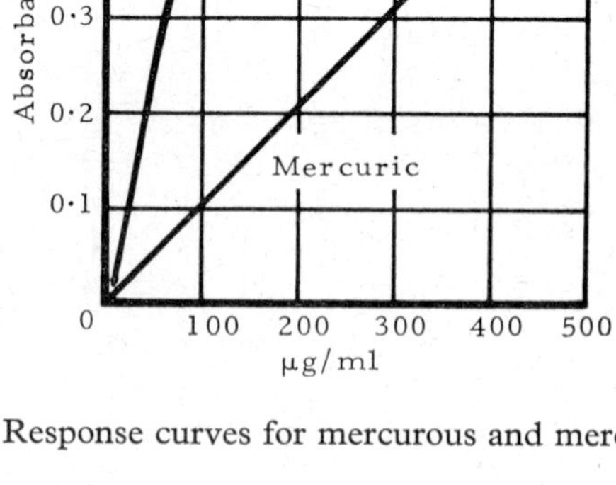

Fig. 4.3 Response curves for mercurous and mercuric salts

10 μg/ml can be reliably distinguished. The sensitivity for 1 per cent absorption is about 5.0 μg/ml. At a scale expansion of $\times$10 the practical

limit of detection for aqueous mercuric solutions is 0·6 μg/ml. Mercurous mercury, it is seen, may be determined over the range 2·0 to 100 μg/ml and differences of 2·0 μg/ml can be discerned. The theoretical sensitivity for 1 per cent absorption is 1 μg/ml, and at a scale expansion of ×10 the practical limit of detection is 0·2 μg/ml in aqueous solution. Above about 100 μg/ml the response curve for mercurous salts shows very pronounced curvature and readings above this concentration are unreliable.

Interferences and Special Characteristics

As stated above, solutions of mercurous salts exhibit a higher absorbance than solutions of mercuric salts containing the same concentration of mercury. Further enhancement to a peak value may be obtained by treating solutions of either mercurous or mercuric salts with an excess of ascorbic acid or stannuous chloride. This phenomena is due to the complete reduction of the mercury to the colloidal elemental form. Due to the rapid aggregation of the colloidal mercury, this procedure can only be utilized to enhance the response for very dilute solutions, and even then the absorption measurement must be made immediately after the addition of the reducing agent.

Heavy excess of the following cations do not affect the absorption response for mercury: Sb, Bi, Cd, Cu, and Zn.

Molybdenum

Wavelength

The 3133 Å line is most frequently used. The line at 3798 Å has also been used by Mostyn and Cunningham.

Flame Systems

Either the luminous air–acetylene or nitrous oxide–acetylene flame may be used. Either system allows the element to be determined in aqueous solution, without scale expansion, over a range from 1·5 to 300 μg/ml.

Sensitivity

Under the most advantageous conditions commercial atomic absorption equipment is capable of yielding a sensitivity for 1 per cent absorption of about 0·6 μg/ml with either flame system.

The limit of detection for an aqueous solution at ×10 scale expansion with either the nitrous oxide–acetylene or air–acetylene flame is better than 0·2 μg/ml.

INTERFERENCES AND SPECIAL CHARACTERISTICS

We find in our laboratory (E.E.L., R.J.R.) that heavy excess of Cd, Cr, Cu, Fe, Pb, Mn, Ni, and Sn have negligible effect upon the absorption of molybdenum when either of the above flame systems is used.

Other workers have reported interference from Cr, Fe, Mn, and Ni, but found that this could be overcome by dosing standards and test solutions with ammonium chloride, up to a level of 2 per cent.

The determination of molybdenum by atomic absorption spectroscopy is, of course, the basis of the indirect amplification procedures for phosphorus, silicon, niobium, and titanium that have been developed by WEST and his school at Imperial College, London.

Nickel

WAVELENGTHS

The most sensitive wavelength is at 2320 Å. This is very close to an ionic line (2316 Å) that can be excited by collision between filler gas and sputtered nickel atoms.

The problem of reducing this unwanted ionic radiation was initially overcome by the development of high intensity lamps. The newer hollow-cathode lamps of conventional design, however, attain the same end simply by restricting the filler gas to a very low pressure. These newer lamps provide a sufficiently clean and intense spectrum for the 2320 Å line to be used with standard production instruments.

In setting to the 2320 Å wavelength, it is essential to distinguish it from the less absorbing line at 2311 Å, which is of only slightly lower intensity. This is accomplished by using narrow monochromator slit widths and approaching from the higher end of the spectrum. 3415 Å and 3525 Å were originally the most commonly used nickel lines and are still valuable for many determinations. Other less absorbing lines exist at 2346, 2525, 3003, 3051 and 3462 Å.

FLAME SYSTEM

Air–acetylene.

SENSITIVITY

At 2320 Å the element can be determined without scale expansion over a range from 0·2 to 25 μg/ml and differences of 0·2 μg/ml can be estimated. Standard commercial instruments are generally capable of a sensitivity for 1 per cent absorption of about 0·1 μg/ml. At a scale expansion of $\times 10$ the limit of detection is about 0·02 μg/ml.

At 3415 Å nickel can be determined without scale expansion over the range 1·0 to 50 μg/ml and differences of about 1·0 μg/ml discerned. The sensitivity for 1 per cent absorption at this wavelength is about 0·5 μg/ml. At a scale expansion of ×10 the limit of detection is 0·05 μg/ml.

INTERFERENCES AND SPECIAL CHARACTERISTICS

Nickel is remarkably free from interference effects from foreign ions, and solutions of the nitrate, sulphate, and chloride that contain equal concentrations of the element give rise to identical responses. The following elements, even when present in heavy excess, do not affect the absorption of nickel: Al, Ca, Cd, Cr, Co, Cu, Fe, Pb, Mg, Mn, Mo, Ag, Na, Sn, W, V, and Zn.

Niobium

Niobium is not an easy metal to determine by classical analytical procedures and it would indeed be cause for elation in the laboratories responsible for its routine estimation if it could be reported that atomic absorption provided a rapid and reliable means for the analysis. Unfortunately this is not so.

The most convincing procedure utilizing atomic absorption is the indirect amplification method developed by WEST and his workers at Imperial College. In principle the method depends upon the formation of molybdoniobophosphoric acid which is extracted into butanol. The phosphomolybdic acid, which is also formed during the reaction, is previously removed by selective extraction into isobutyl acetate.

The molybdenum content of the molybdoniobophosphoric complex is estimated by atomic absorption spectroscopy, and amplification is obtained because eleven molybdenum atoms are associated with every niobium atom. The procedure is of particular value because tantalum does not form a similar complex. The procedure allows niobium to be determined down to a concentration level of about 0·03 μg/ml Nb in the original solution. Heavy excess of Al, Ag, Bi, Be, Ca, Cd, Cr, Co, Cu, Fe, Ni, Mg, Mn, Sb, Ta, V, W, Zn, Zr, SO_4^{2-}, NO_3^-, F^-, and E.D.T.A. do not interfere with the analysis. The presence of titanium causes interference.

The less common Noble Metals

Osmium, Rhenium, and Ruthenium have been determined by atomic absorption spectroscopy.

Briefly, the conditions reported as having been used in the determinations most adaptable to standard commercial equipment are:

Osmium

WAVELENGTHS

2909 Å, also 3059 Å (less sensitive).

FLAME SYSTEM

Nitrous oxide–acetylene.

SENSITIVITY

For 1 per cent absorption, 2 μg/ml.

Rhenium

WAVELENGTH

3461 Å and 3728 Å.

FLAME SYSTEM

nitrous oxide–acetylene.

SENSITIVITY

For 1 per cent absorption, 10 μg/ml.

Ruthenium

WAVELENGTH

3499 Å and 3728 Å.

FLAME SYSTEM

Air–acetylene.

SENSITIVITY

For 1 per cent absorption, 1·5 μg/ml.

Palladium

WAVELENGTHS

3405 Å was the line originally most frequently used. That at 2476 Å is now recognized to be more sensitive. Determinations made at 2448 and 2763 Å have been reported.

Flame System

Early work suggested that the use of a cool air–propane flame would be advantageous for the estimation of palladium. It is now established, though, that the air–acetylene system is preferable.

Sensitivity

At 2476 Å the sensitivity for 1 per cent absorption is about 0·5 μg/ml. At a scale expansion of ×10 the limit of detection in aqueous solution is better than 0·1 μg/ml. Without scale expansion palladium can usefully be determined in aqueous solution over a range from 1·0 to about 125 μg/ml and differences of 1·0 μg/ml reliably estimated.

Interferences and Special Characteristics

The noble metals are remarkably free from cationic interference effects. The absorption of palladium is affected by the presence of excess acid. This effect, though, levels out as the acid concentration rises, and can be overcome by working with solutions that are made 10 per cent V/V with concentrated hydrochloric acid.

Phosphorus

This element can be determined indirectly by atomic absorption spectroscopy by using an indirect amplification procedure.

The principle of the method depends upon the formation and extraction into an organic phase of phosphomolybdic acid $H_3PO_4(MoO_3)_{12}$. The molybdenum is then determined by atomic absorption. Amplification is obtained because 12 molybdenum atoms are associated with every atom of phosphorus. The method allows as little as 0·08 μg/ml of phosphorous in the original test solution to be determined.

The presence of heavy loadings of Al, Sb, As, Au, Bi, Ca, Cd, Co, Cr, Cu, Fe, Ge, Ni, Pb, Mg, Mn, Se, Te, Ti, V, Zn, F^-, E.D.T.A., NO_3^-, and SO_4^{2-}, do not interfere. Tungsten, present in heavy excess in the original test solution, gives rise to low results for phosphorus.

Platinum

Wavelength

The only line of real value to the analyst is that at 2659 Å. When setting to this wavelength, it is advisable to approach from the higher end of the spectrum to ensure that it, and not the less absorbing line at 2647 Å, is selected.

FLAME SYSTEM

Air–acetylene. Air–propane is also suitable for the estimation of platinum, but offers no advantages over air–acetylene.

SENSITIVITY

The sensitivity for 1 per cent absorption attainable with standard atomic absorption equipment is about 3·0 μg/ml. The limit of detection at ×10 scale expansion for an aqueous solution is 0·5 μg/ml.

Without scale expansion platinum can be determined in aqueous solution over the range 6 to 500 μg/ml and differences of about 6 μg/ml can be estimated.

INTERFERENCES AND SPECIAL CHARACTERISTICS

By analogy with the other noble metals it would be expected that the absorption for platinum would be free from cationic interference effects. This has in general been borne out by several workers, but others have reported interferences, which it is stated can be greatly reduced by dosing standard and test solutions with a heavy loading of copper.

Potassium

The determination of the alkali metals by flame emission, is of course, a well established procedure, and this fact, together with the difficulties initially encountered with the production of hollow-cathode lamps for these elements, tended to retard the development of atomic absorption procedures for them, but reliable lamps are now available.

WAVELENGTH

7665 Å is the only line of practical value. In order to make determinations at this wavelength, the instrument must be fitted with a suitable red-end photomultiplier tube.

FLAME SYSTEM

Air–acetylene. Air–propane can also be used. The sensitivity obtained with this flame is similar to that with air–acetylene, but the response curve is distinctly more linear at concentrations above 2 μg/ml.

SENSITIVITY

The sensitivity for 1 per cent absorption is in the region of 0·02 μg/ml. The limit of detection in aqueous solution at ×10 scale expansion is about 0·002 μg/ml.

Without scale expansion the element can be determined in aqueous solution over a range from 0·04 to 5 μg/ml and differences of 0·04 μg/ml can be reliably estimated.

Interferences and Special Characteristics

Potassium is surprisingly free from interference effects from other anions and cations that may be present. Solutions of the chloride, nitrate, and sulphate that contain the same concentrations of potassium give rise to identical responses. The presence of heavy excess of Ca, Li, Mg, and Na do not affect the absorption of potassium. Iron exerts a slight, and phosphate a considerable, depression.

These interferences are overcome by the addition of lanthanum chloride solution.

The Rare Earths

Of this group of elements the following have been determined by atomic absorption spectroscopy: dysprosium, erbium, europium, gadolinium, gallium, hafnium, lanthanum, lutetium, neodymium, proseodymium, samarium, scandium, terbium, thorium, thulium, ytterbium, yttrium.

The need to determine these elements though, is rather infrequent in most laboratories.

Hollow-cathode lamps are expensive for the rare earths, and unfortunately for the rare earth chemist, he is likely to require a different lamp for almost every element of the group. For this reason alone, atomic absorption may not be a very attractive proposition for assaying the rare earths.

It so happens that the elements are amenable to determination by high temperature flame emission spectroscopy. The development of flame spectrophotometers for atomic absorption, fitted with high resolution monochromators and capable of use as emission instruments, is likely to assist here. The technique of separated flame spectroscopy holds out further possibility of reliable, rapid, and inexpensive determinations in this field.

Rhodium

Wavelengths

The most absorbing line is at 3435 Å. Other less absorbing lines occur at 3397, 3503, 3658 and 3692 Å.

FLAME SYSTEM

Air–acetylene.

SENSITIVITY

A sensitivity for 1 per cent absorption of 1 μg/ml is attainable with standard atomic absorption equipment.

The limit of detection in aqueous solution at a scale expansion of ×10 is stated to be 0·1 μg/ml.

Without the use of scale expansion the element can be determined in aqueous solution over a range from 2 to 100 μg/ml and differences of about 2 μg/ml can be discerned.

Rubidium

Hollow-cathode lamps for rubidium are not at the present time very satisfactory in operation, but vapour discharge lamps do provide a convenient radiation source for the element.

WAVELENGTH

7800 Å. In order to make determinations at this wavelength the instrument must be fitted with a suitable red-sensitive photomultiplier.

FLAME SYSTEM

Either the air–hydrogen or air–acetylene flame systems can be used. Very pronounced ionization depression occurs with the air–acetylene system and it is essential to dose standard and test solutions with a heavy excess of potassium (say 2000 μg/ml K) if this flame is used.

SENSITIVITY

In aqueous solution the element can be determined over the range 0·2 to 10 μg/ml and differences of about 0·2 μg/ml can be discerned. The sensitivity for 1 per cent absorption is 0·08 μg/ml. The limit of detection at ×10 scale expansion is 0·01 μg/ml.

INTERFERENCES AND SPECIAL CHARACTERISTICS

Of the elements commonly associated with rubidium, potassium, sodium, lithium and calcium do not affect the absorption in the air–hydrogen flame. Provided the test and standard solutions are adjusted to contain a heavy excess of an alkali metal the same elements do not interfere with the absorption when the air–acetylene system is used.

Selenium

Wavelength

The resonance line at 1960 Å is the most absorbing. There is also a less absorbing line at 2040 Å.

Flame System

The remarks made for arsenic apply also to selenium, namely that all flame systems themselves absorb strongly at wavelengths below 2000 Å. Flames using hydrogen as a fuel are more transparent than air–acetylene at 1961 Å and so are likely to be advantageous for the determination of selenium. Interference due to 'light scatter' is again likely to be troublesome when the solutions contain heavy loadings of foreign materials.

Sensitivity

The sensitivity for 1 per cent absorption is 0·4 μg/ml.

A limit of detection in aqueous solution at $\times 10$ scale expansion of 0·15 μg/ml is possible using the argon–hydrogen flame.

Future Possibilities

West *et al.* have employed atomic fluorescence, using microwave discharge tubes as spectral sources, and have obtained a limit of detection of 0·25 μg/ml at the 2040 Å line. Negligible interference was produced by the presence of excesses of Ag, Al, Co, Cu, Fe, Hg, Mg, NH_4^+, Pb, Sb, Te, and Zn.

The general improvements in hardware that now make atomic absorption measurements possible at the 1960 Å line, and the fact that atomic fluorescence is even more seriously affected by 'light scatter' than is atomic absorption, do not suggest that fluorescence will provide a more reliable procedure for trace estimations of selenium than absorption.

Silicon

Wavelength

Reliable hollow-cathode lamps capable of giving sufficient output to permit realistic determinations for silicon to be made with standard production instruments are now available. The wavelength used is 2516 Å.

FLAME SYSTEM

Nitrous oxide–acetylene. Fuel rich.

SENSITIVITY

The sensitivity for 1 per cent absorption attainable with standard commercial atomic absorption equipment is about 2 μg/ml.

In aqueous solution the limit of detection at a scale expansion of $\times$10 is about 0·3 μg/ml.

Without scale expansion the element can be determined over a range from 4 to 800 μg/ml and increments of about 4 μg can be reliably estimated.

INTERFERENCES AND SPECIAL CHARACTERISTICS

Solutions of sodium meta–silicate and hydrofluosilicic acid that contain the same concentrations of silicon give rise to identical responses. The presence of heavy loadings of the following cations have no effect upon the absorption of silicon: Al, Ca, Fe, Na.

The determination of silicon by an indirect amplification procedure, utilizing the silicomolybdic acid, and allowing silicon to be estimated over a concentration range from 0·08 to 1·2 μg/ml has been described by WEST and his workers. This procedure is remarkably free of interference from foreign ions.

Silver

WAVELENGTH

The line at 3281 Å offers maximum sensitivity. The less absorbing line at 3383 Å is of value for determinations in solutions containing up to about 20 μg/ml Ag.

FLAME SYSTEM

Air–acetylene.

SENSITIVITY

For standard atomic absorption spectrophotometers the sensitivity for 1 per cent absorption can be expected to be about 0·05 μg/ml.

The limit of detection attainable at a scale expansion of $\times$10 is about 0·008 μg/ml.

In aqueous solution, without the use of scale expansion, the element can be determined at the 3281 Å line over the range 0·1 to 10 μg/ml and differences of about 0·1 μg/ml can be reliably estimated.

Interferences and Special Characteristics

Atomic absorption determinations for silver are remarkably free from interference effects and the following anions and cations, commonly found in association with the element, do not affect its absorption, even when present in heavy excess: Al, Ba, Be, Bi, Cd, Ca, Cr, Co, Cu, Fe, La, Pb, Li, Mg, Mn, Hg, Mo, Ni, K, Na, Sr, Sn, Ti, Zn, NH_4^+, CN^-.

Sodium

The determination of sodium in low concentrations is easily carried out by either atomic absorption or emission flame photometry. The latter technique is, of course, a firmly established procedure in most laboratories and this fact has tended to retard the development of absorption procedures for sodium determinations.

Wavelengths

The doublet with peaks at 5890 and 5896 Å provides the only wavelengths of practical value. Hollow-cathode lamps capable of providing sufficiently intense emission for the use of narrow monochromator slit settings are now available, and under these conditions it is possible to resolve the two lines. It is then found that the absorption at 5890 Å is superior to that at 5896 Å.

Flame Systems

Either the air–acetylene or air–propane flame may be used, air–acetylene giving rise to a slightly superior response.

Sensitivity

The sensitivity for 1 per cent absorption is about 0·02 μg/ml. At a scale expansion of ×10 the limit of detection for an aqueous solution is 0·002 μg/ml.

Without scale expansion the element can be estimated in aqueous solution over a range from 0·04 to 5 μg/ml and increments of 0·04 μg/ml can be differentiated.

Interferences and Special Characteristics

Solutions of sodium chloride, nitrate, or sulphate containing equal concentrations of sodium exhibit identical responses.

Heavy loadings of the following cations and anions do not affect the absorption a, Li, Mg, K, PO^{3-}.

Strontium

Wavelengths

The resonance line at 4607 Å is the most absorbing and most frequently used.

The ionic line at 4077·7 Å could be useful for determinations in the range 25–1000 μg/ml Sr.

Flame Systems

The air–acetylene flame system is most commonly employed and using this without scale expansion a sensibly linear response is obtained for aqueous solutions over the range 0·2 to 20 μg/ml Sr. Differences of about 0·2 μg/ml can reliably be estimated.

At the higher temperature of the nitrous oxide–acetylene flame strontium is ionized, so that an aqueous solution of the chloride exhibits much lower absorption than it would in the air–acetylene flame. This lost absorption can be restored by the addition of a heavy excess of potassium ions and under these conditions reliable determinations can be carried out over the range 0·5–40 μg/ml and differences of 0·5 μg/ml can be discerned.

Sensitivity

The sensitivity for 1 per cent attainable with standard atomic absorption equipment, using the air–acetylene flame system, is about 0·1 μg/ml. For the nitrous oxide–acetylene flame the sensitivity for 1 per cent absorption, using aqueous solutions dosed with potassium, is about 0·25 μg/ml. At a scale expansion of $\times 10$ the limit of detection with the air–acetylene flame is 0·02 μg/ml.

Interferences and Special Characteristics

Solutions of the nitrate or chloride that contain equal concentrations of strontium give rise to identical absorptions.

In the air–acetylene flame heavy excess of Ca or Mg does not affect the strontium response. Some enhancement is produced by heavy loadings of Ba, Na, and K. The presence of Al, Fe, and PO_4^{3-} depresses the absorption of strontium but these interferences can be overcome by the addition of lanthanum chloride.

With the nitrous oxide–acetylene flame a much lower response is obtained for aqueous solutions than with air–acetylene, due to much strontium being ionized at the higher temperature. Suppression of this ionization and enhancement of the absorption occurs in the presence of heavy excess of Ba, Ca, K and Na. At the higher temperature of the nitrous oxide–acetylene flame Al is the only element that brings about depression of the strontium absorption, the presence of PO_4^{3-} having no effect.

Tantalum

Like niobium, this element is insensitive to determination by atomic absorption, although one will find lamps advertised, and limits of detection specified, in most instrument makers' catalogues. No indirect procedure for its determination has yet been published.

Tellurium

Wavelength

2143 Å.

Flame System

Air–acetylene.

Sensitivity

The sensitivity for 1 per cent absorption is about 0·5 μg/ml.

In aqueous solution at a scale expansion of ×10 the limit of detection is 0·1 μg/ml.

Special Characteristics

Unlike selenium, tellurium is amenable to determination by atomic absorption spectroscopy. It is also one of the group of elements that West and his school at Imperial College have determined advantageously by atomic fluorescence, using microwave dishcarge tubes as spectral sources. These workers obtained limits of detection of 0·12 μg/ml using fluorescence and 1 μg/ml using atomic absorption.

Negligible interference is caused by excesses of Ag, Al, Co, Cu, Fe, Hg, Mg, NH_4^+, Pb, Se, Sb, and Zn upon either fluorescence or absorption measurements.

Tin

Wavelength

The line at 2246 Å is the most absorbing, and that at 2355 Å slightly less so. The two lines originally most frequently used were 2863 and

2840 Å. In our laboratories (E.E.L.) we find the 2863 Å line to be marginally superior to that at 2840 Å. Other less absorbing wavelengths exist at 2707, 3009 and 3634 Å.

Flame Systems

Tin possesses a refractory oxide and its absorption in the air–acetylene flame is therefore low. An improved response can be obtained by using the much hotter nitrous oxide–acetylene system, but by far the most sensitive conditions are attained with the relatively cool but strongly reducing air–hydrogen flame. Not only is the response enhanced with this latter system, but it is also much more stable. Unfortunately, though, interferences are more likely to occur when the cooler flame is used.

Sensitivity

The essential data for the two lines 2246 and 2863 Å, using the two flame systems are tabulated below.

Interferences and Special Characteristics

The absorption of tin in the air–hydrogen flame, although more stable and pronounced than with the nitrous oxide–acetylene flame, is more prone to interference effects. It could, therefore, be advantageous for the determination of tin in alloys, ores, etc. to utilize the hotter flame system.

Wave-length Å	Flame system	Sensitivity for 1% absorption µg/ml	Limit of detection at ×10 scale expansion µg/ml	Useful range for determination without scale expansion µg/ml	Least increment that can be discerned without scale expansion µg/ml
2246·1	Air–hydrogen	0·4	0·05	1–150	1
2246·1	N_2O–acetylene	2·4	0·3	5–300	5
2863·3	Air–hydrogen	0·8	0·1	2–100	2
2863·3	N_2O–acetylene	4·5	0·6	10–600	10

Air–Hydrogen Flame. The presence of Al and Sb in heavy excess noticeably depresses the absorption response for tin, as does SO_4^{2-}.

The following elements produce very slight depressive effects: Cd, Cu, Fe, Pb, Ni, Na, and Zn.

The presence of an excess of hydrochloric acid also depresses the absorption of tin.

Nitrous Oxide–Acetylene Flame. With this hotter flame system, only antimony, when present in heavy excess, exerts a depressive interference upon tin.

Titanium

Wavelength

The titanium spectrum is rather complex. The line most frequently. used is at 3643 Å. Other less absorbing lines exist at 3654, 3200, 3753 and 3999 Å, to quote only a few.

Flame System

A fuel-rich nitrous oxide–acetylene flame is used.

Sensitivity

The absorption of titanium is drastically enhanced in the presence of 2 per cent HF.

The sensitivity for 1 per cent absorption in this medium is about 3 μg/ml. At a scale expansion of $\times 10$ the limit of detection is about 0·4 μg/ml for an aqueous solution.

In an aqueous solution containing 2 per cent HF titanium can be estimated without scale expansion over the range 6 to 500 μg/ml and differences of about 6 μg/ml can be estimated. Because of the poor sensitivity of this element to determination by atomic absorption, measurements are commonly made on solutions prepared in a medium of $1+1$ isopropyl alcohol and water. In this solvent the useful range for determination without scale expansion is from 3 to 250 μg/ml over which differences of about 3 μg/ml can be detected.

Interferences and Special Characteristics

Titanium is not only insensitive to atomic absorption, but is prone to complex interferences. The presence of potassium dichromate depresses the titanium absorption, whereas the addition of chromic chloride enhances the response. A solution of the potassium titanium oxalate gives rise to an absorption reading identical with that of a solution of titanium chloride containing the same concentration of titanium.

The presence of vanadium also depresses the titanium absorption. Heavy loadings of nickel and iron do not affect the absorption of titanium in the nitrous oxide–acetylene flame.

It is essential to match the acidity of standard and test solutions when the titanium content of alloys is being determined by atomic absorption spectroscopy.

An indirect amplification procedure, capable of a sensitivity for 1 per cent absorption corresponding to 0·013 μg/ml Ti in the aqueous solution, before extraction, has been developed by WEST *et al.* at Imperial College, London. In principle the method depends upon the formation of molybdotitanophosphoric acid, in which eleven atoms of molybdenum are combined with two of titanium.

In brief the details of the procedure are as follows. Molybdotitanophosphoric acid is formed in aqueous solution, made 0·5 M with hydrochloric acid, by the addition of molybdate, phosphate, and potassium alum (to act as a masking agent when fluoride is present in the solution). The excess molybdophosphoric acid is extracted away from the titanium complex into a mixture of chloroform and butanol (4 : 1 by volume). The titanium complex is then extracted into butanol, and the organic phase washed with 0·1 M hydrochloric acid. The concentration of molybdenum in the butanol extract is finally measured by atomic absorption spectroscopy using the nitrous oxide–acetylene flame.

The presence of the following ions, in heavy excess in the aqueous phase from which the titanium complex is extracted, has no effect upon the determination: Ba, Be, Bi, Ca, Cr, Co, Cu, Fe, Pb, Mg, Mn, Ni, Se, Sr, Te, Zn, B, Cl, NO_3^-, SO_4^{2-}. Of the elements that also form heteropoly acids under similar conditions, As and Ge can be removed by volatilization of the chlorides, and Si is eliminated by volatilization of fluosilicic acid during sample preparation.

Tantalum, niobium and zirconium produce positive interferences.

Tungsten

WAVELENGTHS

Absorption can be obtained at the following wavelengths: 4009, the doublet at 2947·4 and 2947·0, 2724, 2681 and 2551 Å. Many other less sensitive lines of the tungsten spectrum have also been examined.

Using an argon-filled hollow-cathode lamp the most advantageous signal-to-noise ratio is obtained at the 4009 Å line. With the newer neon-filled lamps the more sensitive doublet at 2947 Å can be used. The line at 2551 Å shows about the same sensitivity as the 2947 Å doublet at concentrations up to 2000 μg/ml W, but the response curve is less linear at higher concentrations.

Flame System

A fuel-rich nitrous oxide–acetylene flame is required.

Sensitivity

The sensitivity for 1 per cent absorption is about 10 μg/ml at 2551 Å. The limit of detection at $\times 10$ scale expansion is stated to be 5 μg/ml for an aqueous solution.

In aqueous medium the element can be determined at 2551 or 2947 Å without scale expansion over a range from 20 to 4000 μg/ml and differences of about 20 μg/ml can be estimated. Because of the poor sensitivity of tungsten towards determination by atomic absorption it is usual to operate with solutions prepared in a medium of 1+1 isopropyl alcohol and water. Under these conditions the element can be determined without scale expansion over the range 5 to 1000 μg/ml and differences of 5 μg/ml can be detected.

Interferences and Special Characteristics

The presence of heavy loadings of vanadium slightly depresses the absorption of tungsten. No interference is caused by the presence of Cr, Ni, or Mo.

Vanadium

Wavelength

The unresolved triplet with peaks at 3183, 3184 and 3185 Å gives rise to the most sensitive conditions. Other less absorbing lines occur at 3066 Å and the 4385–4390 Å doublet.

Flame System

Nitrous oxide–acetylene.

Sensitivity

Sensitivity for 1 per cent absorption is about 2·0 μg/ml. The limit of detection for aqueous solutions at $\times 10$ scale expansion is approximately 0·2 μg/ml.

Without scale expansion the element can be estimated in aqueous solution over a range from 4 to 750 μg/ml and differences of 4 μg/ml can be detected.

INTERFERENCES AND SPECIAL CHARACTERISTICS

The absorption of vanadium is unaffected by heavy excesses of Cr, Co, Fe, Mn, Mo, Ni, or W.

Uranium and Zirconium

Neither of these elements is sensitive to determination by atomic absorption and although lamps are listed in makers catalogues the use of the technique for their determination cannot be recommended.

Zinc

WAVELENGTH

Estimations are always carried out at the 2139 Å line.

Response curves using two different lamps are shown in Fig. 4.4. The most sensitive conditions were obtained with a lamp containing a pure zinc cathode. The second curve was obtained with a lamp having a brass cathode. From these curves it can be seen that without scale expansion zinc can be determined in aqueous solution, with the pure zinc hollow-cathode lamp, over a range from 0·04 to 2 μg/ml and that differences of 0·04 μg/ml can be detected. With the brass cathode lamp determinations can be made over a range from 0·1 to 5 μg/ml and differences of 0·1 μg/ml estimated.

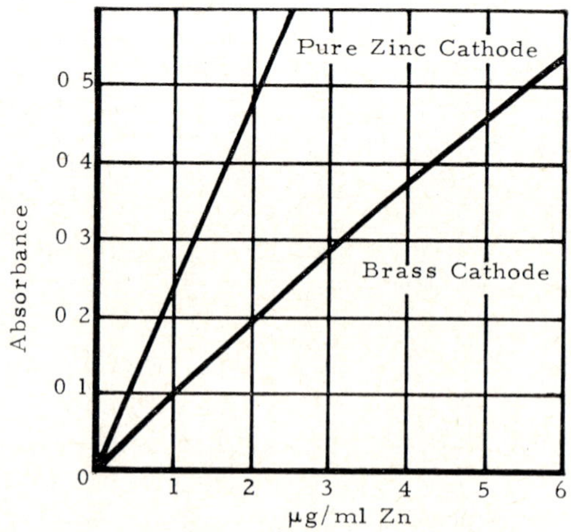

Fig. 4.4. Response curves for zinc

Flame System

The air–acetylene flame is most commonly used. The air–propane flame gives a slightly enhanced, and the air–hydrogen flame a slightly lower, response. The response with air–hydrogen is very stable and this flame system is therefore particularly suitable for scale expansion. Zinc may also be determined with the nitrous oxide–acetylene flame, which gives rise to a lower sensitivity and allows the element to be estimated over a range from 1 to 40 μg/ml. The sensitivity with this flame system is not improved by the addition of a heavy excess of potassium ions.

Sensitivity

The sensitivity for 1 per cent absorption for the air–acetylene system is 0·02 μg/ml under the most favourable conditions. The limit of detection for an aqueous solution at a scale expansion of ×10 is 0·004 μg/ml. Precise data for different types of lamps are given in Table IV-1.

Interferences and Special Characteristics

Using the air–acetylene flame, which is the one most commonly employed, solutions of the chloride, nitrate, or sulphate that contain equal concentrations of zinc give rise to identical absorptions. The following substances do not affect the zinc response, even when present in heavy excess: Al, Cd, Ca, Cu, Fe, Pb, Mn, Mg, Ni, Ag, Na, Sn, and PO_4^{3-}.

Chap. IV References

Aluminium

1. Amos, M. D., and Willis, J. B., Use of high temperature premixed flames in atomic absorption spectroscopy. *Spectrochim. Acta* 22 (1966) 1325.

Antimony

2. Mostyn, R. A., and Cunningham, A. F., The determination of antimony by atomic absorption spectrometry. *Anal Chem.* 39 (1967) 433.
3. Meranger, J. C., and Somers, E., Determination of antimony in titanium dioxide by atomic-absorption spectrophotometry. *Analyst* 93 (1968) 799.

ARSENIC

4. DAGNALL, R. M., THOMPSON, K. C., and WEST, T. S., Studies in atomic-fluorescence spectroscopy—III. Microwave-excited electrodeless discharge tubes as spectral sources for atomic-fluorescence and atomic-absorption spectroscopy. *Talanta* 14 (1967) 551.

5. DAGNALL, R. M., THOMPSON, K. C., and WEST, T. S., Studies in atomic-fluorescence spectroscopy—VII. Fluorescence and analytical characteristics of arsenic with microwave-excited electrodeless discharge tubes as the spectral source. *Talanta* 15 (1968) 677.

BARIUM

6. KIORTYOHANN, S. R., and PICKETT, E. E., Spectral interferences in atomic absorption spectrometry. *Anal Chem.* 38 (1966) 585.

CALCIUM

7. DAVID, D. J., The determination of calcium in plant material by atomic absorption spectrophotometry. *Analyst* 84 (1959) 536.
8. WILLIS, J. B., The determination of calcium and magnesium in urine by atomic absorption spectroscopy. *Anal. Chem.* 33 (1961) 556.
9. WILLIS, J. B., The determination of metals in blood serum by atomic absorption spectroscopy—I. Calcium. *Spectrochim. Acta* 16 (1960) 259.
10. DICKSON, R. E., and JOHNSON, C. M., Interferences associated with the determination of calcium by atomic absorption. *Appl. Spectry.* 20 (1966) 214
11. BENTLEY, E. M., and LEE, G. F., The determination of calcium in natural water by atomic absorption spectroscopy. *Environmental Sience and Tech.* 1 (1967) 721.

CHROMIUM

12. FELDMAN, F. J., and PURDY, W. C., The atomic absorption spectroscopy of chromium. *Anal. Chim. Acta* 33 (1965) 273.
13. GIAMMARISE, A., The use of ammonium chloride in analyses of chromium samples containing iron. *Atomic Abs. Newsletter* 5 (1966) 113.

COBALT

14. HARRISON, W. W., Factors affecting the selection of a cobalt analysis line for atomic absorption spectrometry. *Anal. Chem.* 37 (1965) 1168.

COPPER

15. ALLAN, J. E., The determination of copper by atomic absorption spectrophotometry. *Spectrochim. Acta* 17 (1961) 459
16. KHALIFA, H., SVEHLA, G., and ERDEY, L., Precision of the determination of copper and gold by atomic absorption spectrophotometry. *Talanta* 12 (1965) 703.

GALLIUM

17. MULFORD, C. E., Gallium and indium determinations by atomic absorption. *At. Abs. Newsletter* 5 (1966) 63.

GOLD

18. TINDALL, F. M., Silver and gold assay by atomic absorption spectrophotometry. *At. Abs. Newsletter* 4 (1965) 339.
19. SIMMONS, E. C., Gold assay by atomic absorption spectrophotometry. *At. Abs. Newsletter* 4 (1965) 281.
20. TINDALL, F. M., Notes on silver and gold assay by atomic absorption. *At. Abs. Newsletter* 5 (1966) 140.

INDIUM

21. MULFORD, C. E., Gallium and indium determinations by atomic absorption. *At. Abs. Newsletter* 5 (1966) 63.

IRIDIUM

22. MULFORD, C. E., Iridium absorption. *At. Abs. Newsletter* 5 (1966) 63.
23. MANNING, D. C., and FERNADEZ, F., Iridium determination by atomic absorption. *At. Abs. Newsletter* 6 (1967) 15.

LEAD

24. DAGNALL, R. M., and WEST, T. S., Observations on the atomic absorption spectroscopy of lead in aqueous solution, in organic extracts and in gasoline. *Talanta* 11 (1964) 1553.
25. BERMAN, E., The determination of lead in blood and urine by atomic absorption spectrophotometry. *At. Abs. Newsletter* 3 (1964) 111.
26. CHAKRABARTI, C. L., ROBINSON, J. W., and WEST, P. W., The atomic absorption spectroscopy of lead. *Anal. Chim. Acta* 34 (1966) 269.

MAGNESIUM

27. ALLAN, J. E., Atomic absorption spectrophotometry with special reference to the determination of magnesium. *Analyst* 83 (1958) 466.
28. DAVID, D. J., The determination of exchangeable sodium, potassium, calcium and magnesium in soils by atomic absorption spectrophotometry. *Analyst* 85 (1960) 495.
29. WILLIS, J. B., The determination of metals in blood serum by atomic absorption spectroscopy—II. Magnesium. *Spectrochim. Acta* 16 (1960) 273.
30. HALLS, D. J., and TOWNSHEND, A., Some interferences in the atomic absorption spectrophotometry of Mg. *Anal. Chim. Acta* 36 (1966) 278.
31. FIRMAN, R. J., Interference caused by iron and alkalis on the determination of magnesium. *Spectrochim. Acta* 21 (1965) 341.

MERCURY

32. HINGLE, D. N., KIRKBRIGHT, G. F., and WEST, T. S., Some observations on the determination of mercury by atomic absorption spectroscopy in an air–acetylene flame. *Analyst* 92 (1967) 759.

MOLYBDENUM

33. KIRKBRIGHT, G. F., SMITH, A. M., and WEST, T. S., Rapid determination of molybdenum in alloy steels by atomic absorption spectroscopy in a nitrous oxide–acetylene flame. *Analyst* 91 (1966) 700.
34. MOSTYN, R. A., and CUNNINGHAM, A. F., Determination of molybdenum in ferrous alloys by atomic absorption spectrometry. *Anal. Chem.* 38 (1966) 121.
35. DAVID, D. J., The suppression of some interferences in the determination of molybdenum by atomic absorption spectroscopy in an air–acetylene flame. *Analyst* 93 (1968) 79.

NIOBIUM

36. KIRKBRIGHT, G. F., SMITH, A. M., and WEST, T. S., An indirect amplification procedure for the determination of niobium by atomic absorption spectroscopy. *Analyst* 93 (1968) 292.

LESS COMMON NOBLE METALS

37. LOCKYER, R., and HAMES, G. E., Quantitative determination of some noble metals by atomic absorption spectroscopy. *Analyst* 84 (1959) 385.

38. STRASHEIM, A., and WESSELS, G. J., The atomic absorption determination of some noble metals. *Appl. Spectroscopy* 17 (1963) 65.

39. SCHRENK, W. G., LEHMAN D. A., and NENFELD, L., Atomic absorpton characteristics of rhenium. *Appl. Spectroscopy* 20 (1966) 389.

PALLADIUM

40. LOCKYER, R., and HAMES, G. E., Quantitative determination of some noble metals by atomic absorption spectroscopy. *Analyst* 84 (1959) 385.

41. STRASHEIM, A., and WESSELS, G. J., The atomic absorption determination of some noble metals. *Appl. Spectroscopy*, 17 (1963) 65.

42. ERINC, G., and MAGEE, R. J., The determination of palladium by atomic absorption spectroscopy. *Anal. Chim. Acta* 31 (1964) 197.

PHOSPHORUS

43. KIRKBRIGHT, G. F., SMITH, A. M., and WEST, T. S., An indirect sequential determination of phosphorus and silicon by atomic absorption spectrophotometry. *Analyst* 92 (1967) 411.

PLATINUM

44. LOCKYER, R., and HAMES, G. E., Quantitative determination of some noble metals by atomic absorption spectroscopy. *Analyst* 84 (1959) 385.

45. STRASHEIM, A., and WESSELS, G. J., The atomic absorption determination of some noble metals. *Appl. Spectroscopy* 17 (1963) 65.

THE RARE EARTHS

46. KIRKBRIGHT, G. F., SEMB, A., and WEST, T. S., Spectroscopy on separated flames—III. Use of the separated nitrous oxide–acetylene flame in thermal emission spectroscopy. *Talanta* 15 (1968) 441.

47. KRIEGE, O. H., and WELCHER, G. G., Determination of scandium by atomic absorption. *Talanta* 15 (1968) 781.

48. YIU-KEE CHAU, and PUI-YUEN WONG. Determination of scandium in sea-water by atomic-absorption spectroscopy. *Talanta* 15 (1968) 867.

RHODIUM

49. LOCKYER, R., and HAMES, G. E., Quantitative determination of some noble metals by atomic absorption spectroscopy. *Analyst* 84 (1959) 385.

50. STRASHEIM, A., and WESSELS, G. J., The atomic absorption determination of some noble metals. *Appl. Spectroscopy* 17 (1963) 65.

51. HENEAGE, P., A brief study of rhodium absorption. *At. Abs. Newsletter* 5 (1966) 64.

52. DEILY, J. R., The determination of rhodium in organic solutions by atomic absorption spectrometry. *At. Abs. Newsletter* 6 (1967) 66.

SELENIUM

53. DAGNALL, R. M., THOMPSON, K. C., and WEST, T. S., Studies in atomic fluorescence spectroscopy—IV. The atomic-fluorescence spectroscopic determination of selenium and tellurium. *Talanta* 14 (1967) 557.

SILICON

54. KIRKBRIGHT, G. F., SMITH, A. M., and WEST, T. S., An indirect sequential determination of phosphorous and silicon by atomic-absorption spectrophotometry. *Analyst* 92 (1967) 411.

55. PRICE, W. J., and ROOS, J. T. H., The determination of silicon by atomic-absorption spectrophotometry with particular reference to steel, cast iron, aluminium alloys and cement. *Analyst* 93 (1968) 709.

SILVER

56. BELCHER, R., DAGNALL, R. M., and WEST, T. S., Examination of the atomic absorption spectroscopy of silver. *Talanta* 11 (1964) 1257.

57. TINDALL, F. M., Silver and gold assay by atomic absorption spectrophotometry. *At. Abs. Newsletter* 4 (1965) 339.

58. TINDALL, F. M., Notes on silver and gold assay by atomic absorption. *At. Abs. Newsletter* 5 (1966) 140.

TELLURIUM

59. DAGNALL, R. M., THOMPSON, K. C., and WEST, T. S., Studies in atomic fluorescence spectroscopy—IV. The atomic fluorescence spectroscopic determination of selenium and tellurium. *Talanta* 14 (1967) 557.

TIN

60. AGAZZI, E. J., Determination of tin in hydrogen peroxide solutions by atomic absorption spectrometry. *Anal. Chem.* 37 (1965) 364.

61. CAPACHO-DELGADO, L., and MANNING, D. C., Determination of tin by atomic absorption spectroscopy. *Spectrochim. Acta* 22 (1966) 1505.

TITANIUM

62. KIRKBRIGHT, G. F., SMITH, A. M., WEST, T. S., and WOOD, R., An indirect amplification procedure for the determination of titanium by atomic absorption spectroscopy. *Analyst* 94 (1969) 754.

VANADIUM

63. AMOS, M. D., and WILLIS, J. B., Use of high-temperature premixed flames in atomic absorption spectroscopy. *Spectrochim Acta.* 22 (1966) 1325.

64. SACHDEV, S. L., ROBINSON, J. W., and WEST, P. W., Determination of vanadium by atomic absorption spectrophotometry. *Anal. Chim. Acta* 37 (1967) 12.

V

Applications

Determination of Metallic Elements in Biological Materials and Organic Substances generally

THE value of atomic absorption spectroscopy for the determination of trace metals after extraction from biological materials is becoming increasingly accepted. The main problems associated with such analyses are (i) the quantitative extraction of the metallic species from the organic matrix, as rapidly as possible, (ii) the avoidance of mechanical losses, (iii) the avoidance of contamination from reagents.

Digestion Procedures

Dry ashing of biological materials is still practiced, but many authorities condemn this as being too vulnerable to mechanical losses in the smokes evolved.

Wet ashing, with mixtures of perchloric, nitric, and sulphuric acids is also widely practised and in the hands of an experienced operator gives reliable results.

A procedure that has recently been developed[(30)], and that possesses the advantages of speed of attack and complete destruction of the carbonaceous material, is the use of 50 per cent hydrogen peroxide. This has been employed in the E.E.L. laboratories in combination with concentrated sulphuric acid, as recommended by the authors, and also in combination with concentrated nitric acid, with very satisfactory results for the digestions of blood, liver, meat products, wood, leaves, fruit, tallow, soap, and animal feeds.

It is not, of course, valid to describe one standard procedure that will be suitable for the extraction of every element from every substance. The quantities of hydrogen peroxide and acid required will vary with the nature of the material to be attacked, and with the sample size which itself will depend on the trace metal concentration in the sample. The decision whether to use nitric or sulphuric acid in conjunction with the hydrogen peroxide will depend on the nature of the trace metal; e.g. sulphuric could be unsuitable for the extraction of lead since lead sulphate is insoluble.

Levels of determination. The limits of detection in aqueous solution for the elements that can be determined by atomic absorption are stated in Table IV-1, in Chapter IV.

For most determinations of this nature it will be necessary to use scale expansion, and it is an advantage if the expanded signal can be fed to a chart recorder.

Concentration Procedures. If the extract from the digested biological material contains the trace metals at a concentration still too low for estimation, even when scale expansion is utilized, it will be necessary to resort to extractive concentration. A convenient and generally applicable procedure is described in Chapter III (method 5).

General Applications in the Food Industry

In the food, brewing, and mineral water industries there is an insistent demand for accurate, reproducible and rapid analyses of trace metals. It is always advantageous if such analyses can be performed directly upon a process liquor (e.g. a brew just prior to bottling or a random sample from a batch of cans) without further concentration or extraction into an organic phase. For solid foodstuffs, the most convenient procedure is to extract the trace metals by digestion as outlined above with dilute acid or with 50 per cent hydrogen peroxide, and to use the aqueous extract for the determination. Vegetable oils may be similarly treated, and Fig. 5.1 shows, as an example, the determination of nickel in such digests.

Copper and zinc are perhaps the most common toxic elements of interest to the food analyst, while the estimation of iron is often necessary, particularly in the brewing and distilling industries for reasons of taste and colour control. The work of our own laboratory (E.E.L.) has centred around the trace estimations of these three elements, but the procedures outlined above are applicable to trace metals generally.

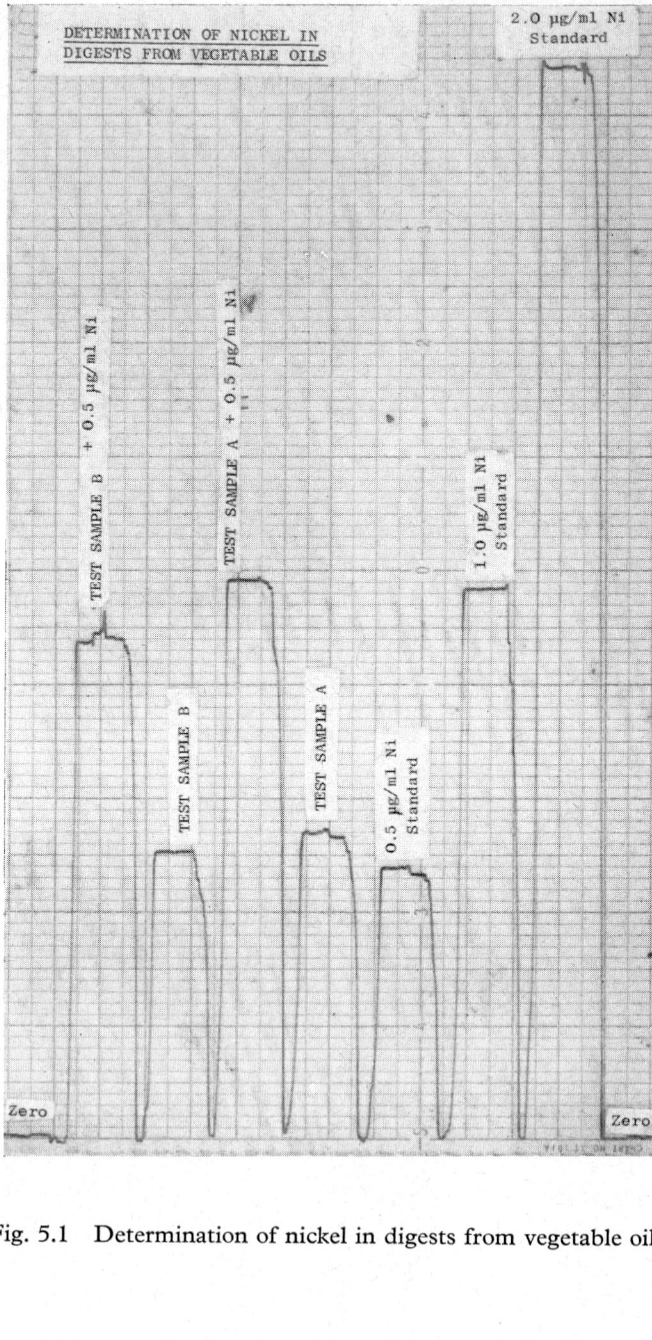

Fig. 5.1 Determination of nickel in digests from vegetable oils

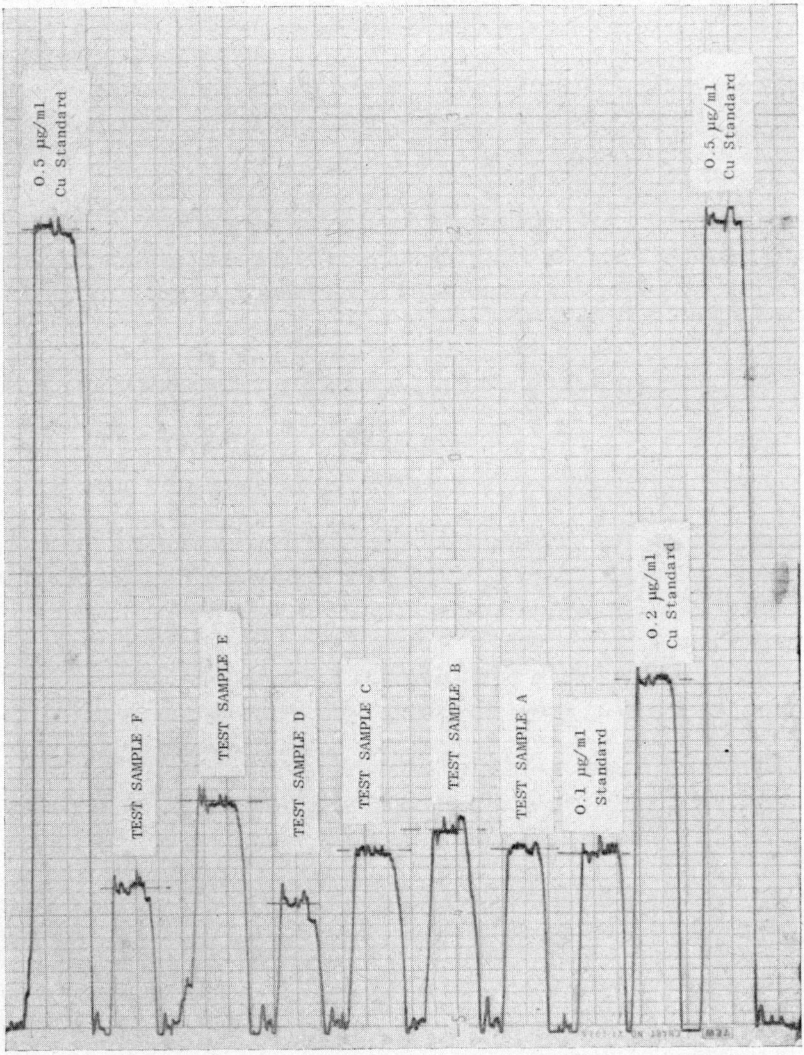

Fig. 5.2 Direct determination of copper in beers

Figure 5.2 shows results for the direct determination of copper in beers. The two important nutritional elements calcium and magnesium are very easily determined by the procedure, but for their determination it is necessary to add lanthanum chloride to the test and standard solutions to overcome the effects of the phosphate suppression.

Applications in clinical Chemistry

THE DETERMINATION OF CALCIUM, MAGNESIUM, SODIUM, AND POTASSIUM IN BLOOD SERUM

Calcium and Magnesium. The determination of calcium and magnesium in blood serum is of importance in diagnosing many pathological conditions, such as renal and liver diseases, hyperparathyroidism, diabetes, maladsorption syndrome and primary aldosterionism. Atomic absorption provides a reliable, reproducible, and rapid procedure for these determinations, and many authoritative workers feel it will eventually be accepted as a referee method.

The procedure is exceedingly simple, sample preparation being limited to diluting the serum 10, 20, or even 50 times, in the presence of lanthanum chloride which overcomes the possible under-estimation of calcium due to phosphate suppression. The test solutions are then aspirated to the atomic absorption spectrophotometer and the responses obtained are compared with those noted for aqueous standard solutions, containing lanthanum chloride at the same level as the test solutions, and with calcium and magnesium contents covering the anticipated ranges in the diluted sera. A very exhaustive series of trials carried out in the E.E.L. laboratories failed to detect any evidence of protein interference when this procedure was used.

A second series of trials indicated that with a normal production instrument, operated without scale expansion, the standard deviation for calcium determinations was equivalent 0·01 mg/100 ml. At a dilution of twenty times, this led to an accuracy of $\pm 0{\cdot}2$ mg/100 ml on the serum, which is well within the range of variation acceptable to clinicians.

By resorting to zero-suppression followed by scale expansion, it is possible to work over the range 0·4–0·6 μg/100 ml and to obtain a standard deviation of 0·001 μg/100 ml, leading to a precision of 0·02 μg/100 ml in the original serum.

Very severe depression of calcium absorption was observed to occur in solutions that were more than 1N with respect to hydrochloric acid. This observation led us to suspect that differences hitherto reported between samples of serum that had been deproteinized and those that had been simply diluted, and which had been attributed to 'protein enhancement', could have been due to *depressed* results being obtained from the deproteinized test solutions, due to their higher acidity.

It is essential that the acid concentration in the test solutions should be less than 0·5N.

It is of advantage for this determination to use a burner with a wide slot, designed to be less susceptible to clogging when solutions containing high concentrations of dissolved solids are aspirated to it.

E.D.T.A. may be used as an interference suppressant in place of lanthanum chloride, and of course reagents that are low in calcium and magnesium content must be employed. Essentially the same procedure is suitable for the estimation of calcium and magnesium in urine.

Sodium and Potassium. The determination of these two elements in biological fluids by flame *emission* spectroscopy is a firmly established procedure and the versatility of most atomic absorption spectrophotometers is well illustrated by the fact that they can be adapted, by the mere flick of a switch, to operate in the emission mode. By emission, both elements can be reliably determined in sera at dilutions of 50 or 100 : 1.

Reliable hollow-cathode lamps are now available for both sodium and potassium, and by atomic absorption sodium can be determined in serum at dilutions between 200 and 500 times, and potassium at dilutions between 50 and 100 times.

For all four elements (calcium, magnesium, sodium, and potassium) exhaustive trials in our own laboratories have failed to detect any serious interference effects due to the presence of protein in the test solutions. A further prolonged series of trials carried out on control sera of all types and makes indicated that the results obtained by atomic absorption were always well within the allowable variations when these were specified by the manufacturers.

Repeatability (perhaps the most important factor to the clinician) is of a high order for the four elements, and the instrumental uncertainties (noise levels) are very low for magnesium and potassium and slightly higher, but still small, for calcium and sodium.

The Determination of Iron in Blood Serum

The determination of iron in blood serum is subject to two controlling factors, namely the very small amount in the serum and the fact that only about 2 ml of serum are likely to be available for test.

Until recently the procedures employed for this estimation by atomic absorption resorted to attack of the serum with trichloroacetic acid, separation of the solution from the solid debris, and complexing the iron and concentrative extraction into an organic solvent. Such procedures offered no advantages in simplicity, speed of operation and accuracy over the established colorimetric methods.

The dramatic improvements in lamp technology that have taken place during the last two years, together with the use of scale expansion and signal integration, now make it possible to perform this estimation upon a sample of serum simply diluted 1+1 with deionized water. Such a procedure is obviously much more rapid and free from handling errors than one involving a series of preparative steps. A standard deviation of about 1 μg/100 ml is calculated for the estimation, leading to a reliability of 2 μg/100 ml in the undiluted serum.

The Determination of Lead in Blood and Urine

An analysis that many workers feel lends itself to the application of atomic absorption is the determination of lead in blood. This is at present most frequently carried out by various modifications of the method developed by Eleanor Berman[(10)]. The determination is complicated by the fact that the amount of lead in blood can vary over a fairly wide range (up to 50 μg per 100 ml for normal healthy people, and well above this level in cases of severe poisoning) and that with children, who form a fairly high percentage of patients, the amount of blood that can be taken is limited to about 5 to 10 ml once a week. Adults can afford to donate 20 to 30 ml once a week.

The essential principles of the method are:

(a) The precipitation of blood proteins by digestion with trichloroacetic acid (Piper and Higgins[(33)] recommend the use of sulphuric-nitric-perchloric mixture).

(b) The formation of a chelated lead complex by the addition of ammonium pyrrolidine dithiocarbamate.

(c) The extraction of this complex into a medium of isobutyl methyl ketone, which can then be sprayed to the atomic absorption spectrophotometer.

For urine, the sample need only be acidified, treated with the chelating agent and extracted into isobutyl methyl ketone. The method allows lead to be estimated to an accuracy of about 0·02 μg/ml, in the organic extract, leading to an accuracy of about 0·04 μg/ml (4 μg per 100 ml) in whole blood.

A rapid determination capable of a precision of about $\pm$16 μg Pb per 100 ml in the whole blood is possible by digesting the sample with a solution of trichloroacetic acid in 1+1 industrial methylated spirits +water, centrifuging off the solid debris and spraying the solution to the spectrophotometer.

As with the iron determination, scale expansion is necessary, and it can confidently be predicted that with improved lamps atomic absorption will, in the not too distant future, provide a rapid and reliable method for this analysis, suitable for adoption by the smaller laboratory.

The early work on this determination was carried out at the 2833 Å line, but improvements in lamp design now allow the more sensitive 2170 Å line to be used with standard instruments. It was hoped that the better sensitivity attainable would enable the estimation to be made directly upon a urine sample, or upon the clear liquor from a sample of blood digested with 50 per cent H_2O_2 and nitric acid, by comparison against straight aqueous lead standards. This simple procedure cannot, unfortunately, be adopted, since very considerable enhancement attributed to light scatter, occurs at 2170 Å due to the sodium content of the test samples. This effect gives rise to erroneously high results if simple aqueous lead standards are used. Correct results can be obtained by adding sodium to the lead standards so that their composition approximately matches that of the test samples, and by using a blank that contains sodium to the same extent.

A proposed method that should allow simple aqueous lead standards and a water blank to be used is that of making the measurement at 2170 Å and then reading the apparent absorption of the test solutions (i.e. the loss of light due to scatter) at the non-absorbing 2204 Å line. By subtracting this apparent absorption from the value read at 2170 Å, the true absorption due to the lead content of the samples should be obtained.

The Determination of Copper in Liver, etc.

In cases of Wilson's disease, copper accumulates in this organ and its determination can be of value to pathologists. The determination is a simple one since copper is a sensitive element to determination by atomic absorption spectroscopy. Even in a normal liver there is between 4·16 and 16·9 μg per g of copper (a considerable loading) and the weight of sample available is usually fairly large. The main problem resolves itself into obtaining an efficient method for extracting the copper, and it has been established that this can be rapidly accomplished with a mixture of sulphuric and nitric acids plus some hydrogen peroxide, or with a mixture of 2 N hydrochloric acid and hydrogen peroxide. The use of 50% hydrogen peroxide gives an equally reliable and even more rapid extraction. The extract from the digest is separated from any solid debris (usually very little), adjusted to 100 ml and used for the determination.

THE DETERMINATION OF GOLD IN SERUM AND URINE

This estimation is of importance in cases of poisoning that have arisen during the treatment of rheumatic disorders by gold injections. It is an example of the type of analysis that is rapidly and reliably made by atomic absorption spectroscopy, and which would be difficult to perform by any other procedure.

Metallurgical Analysis, including Plating Solutions

Because it is possible to determine most metallic elements in concentrations ranging from trace to macro quantities, and because of the general freedom from interferences between these elements, atomic absorption spectroscopy is ideally suited to metallurgical analysis. Indeed, this branch of analytical chemistry was one of the first in which atomic absorption found application.

GENERAL PROCEDURE

The most general, reliable, and convincing procedure used by the atomic absorption spectroscopist for the analysis of alloys is to prepare a range of standards that approximately match in composition the anticipated test solution that will be obtained from the alloy to be analysed. A typical example of this type of procedure is given in Chapter III for a duralumin alloy. The general procedure described has been successfully used for the determination of a very wide range of alloys. We have, in our own laboratories (E.E.L.), established its suitability for the assays listed below, which indicate not only the large number of elements of metallurgical interest that can be determined, but also the wide range of concentrations over which these determinations can be carried out.

1. Co, Al, Ni, Cu, Pb, Mn in steels
2. Al, Co, Fe in high nickel alloys
3. Cu, Sb, Sn in white metals
4. Mg and Zn in cadmium metal
5. Zn, Pb, Ni, Fe, Mg in high copper alloys
6. Cu, Pb, Ni, Fe, Cd, Zn, Mg, Mn in high aluminium alloys
7. Al in brass
8. Cu, Zn, Mn in magnesium alloys
9. Fe, Ca, Al, Mg in iron ore sinter.

INTERFERENCE-FREE DETERMINATIONS

When it is required to develop an atomic absorption procedure for large numbers of similar analyses e.g. the routine control of an alloy (particularly when only a few of the elements present in it are to be estimated) it is always worthwhile examining the possibility of using simple standard solutions. An example of this approach to a metallurgical analysis is given in Chapter III for the estimation of traces of copper, cadmium, and zinc in a tin-lead solder. The major problem here is that even the purest lead and tin salts (readily available to the chemist) contain substantial quantities of cadmium and zinc, so that the preparation of matching standards that contain the two major constituents of the alloy is impracticable.

The following procedure, which utilizes the method of additions, can in this case (where the low concentrations ensure linear response curves) conveniently be used to establish whether standards containing only copper, cadmium, and zinc will allow the determination to be carried out with sufficient accuracy. The trace metals are usually present in the solders at the levels shown below, which when 2 g are dissolved, and the resulting solution adjusted to 100 ml, give rise to the concentrations indicated in the test solution.

Element	% in solder	μg/ml in test solution
Cu	0·01–0·03	2–6
Cd	0·001	0·2
Zn	0·001–0·002	0·2–0·4

Standard solutions would be prepared as follows:

Standard solution I (μg/ml)	Standard solution II (μg/ml)	Standard solution III (μg/ml)
Cu 1·0	Cu 4·0	Cu 8·0
Cd 0·1	Cd 0·2	Cd 0·5
Zn 0·2	Zn 0·4	Zn 0·8

Test solutions of the alloy(s) would then be prepared by dissolution in either hydrochloric or nitric acid, followed by filtration from the precipitated lead chloride or metastannic acid. The precipitate would be washed with a hot 0·1 N solution of the relevant acid and the washings added to the filtrate. Test solution *A* would contain 2 g of solder, dissolved, and the volume adjusted to 100 ml. Test solution *B* would contain 2 g of solder, dissolved, and made up to 100 ml as above, but with the further additions of 200 μg of copper, 100 μg of cadmium, and 100 μg of zinc, so that the concentrations of these three elements in it would be in excess of those in solution *A* by 2 μg/ml copper, 1 μg/ml cadmium, and 1 μg/ml zinc.

The standard and test solutions would now be aspirated to the atomic absorption spectrophotometer and the results obtained for the concentrations of the three elements, as derived from calibration curves constructed from the responses of the standard solutions, compared with those obtained from solutions *A* and *B* by the method of additions. If these two sets of results were identical it would be concluded that the use of simple standards was valid for the determination.

Interference Effects

Although interference of one metal upon another, or of excess acid upon a metal's absorption, does occasionally occur, atomic absorption spectroscopy is in general remarkably free from such effects. An indication of the interferences to be expected between metals is given in Chapter IV, which deals with the characteristics of the elements.

In some cases the interference of one element upon another is enhanced with increasing acid concentration, even though excess acid by itself exerts no effect upon the element's absorption. This effect is exemplified by the interference of aluminium upon magnesium in the presence of excess hydrochloric acid.

Perhaps the most inconvenient and difficult interference to overcome among the assays commonly demanded of the metallurgical chemist is that of iron upon chromium. This is strongly dependent upon the acid concentration of the solution, and strict attention must be paid to the preparation of the test and standard solutions in order to overcome it. Chromium is perhaps the best example of an element more prone to inter-element interferences than would be expected. For this reason two procedures that can be employed for this determination are outlined below. Neither of these is absolutely infallible for the determination of chromium in every class of steel. They have been used with some success for various estimations in the E.E.L. laboratory and are described here only as a rough guide from which a reliable procedure might be developed.

The Determination of Chromium in Steel

Method I. Using the air–acetylene flame the presence of iron strongly depresses the absorption of chromium, and this depression is itself dependent upon the acidity of the solution. For these reasons the determination of chromium in steel is not so straightforward as most metallurgical estimations. The chromium content of most common steels varies from about 1 per cent to 19 per cent. The optimum concentration range for chromium determinations with the air–acetylene system in aqueous solution is between 5 and 20 μg/ml, and the sample size is selected so that the test solutions will contain chromium at this concentration.

A suitably-sized sample of the steel (1 g for a low-chromium steel; 0·2 g for a high chromium steel) is weighed out and dissolved by warming over a low flame with 1+4 concentrated sulphuric acid+water (30 ml is a suitable volume for a 1 g sample). When the reaction has nearly ceased, 10 drops of concentrated nitric acid are added to oxidise carbon. High chrome steels may not be completely dissolved by this procedure, a small metallic residue sometimes remaining. If this situation arises, the solution is poured off from the residue, which is then dissolved by the addition of a few drops of concentrated nitric and hydrochloric acids and added to the main solution. The whole solution is adjusted to a convenient volume (250 ml or 500 ml) to produce a master sample solution.

A suitably-sized aliquot of the master test solution is pipetted to a beaker. Potassium hydroxide (10 per cent solution) is added dropwise with continual stirring until the aliquot is just alkaline, which is then reacidified with the minimum quantity of hydrochloric acid. This solution is transferred to a suitable volumetric flask, sufficient lanthanum chloride solution added to give the final test solution a loading of 6500 μg/ml lanthanum, and adjusted to the mark with deionized water. The chromium content of the test solution is determined by aspirating to an atomic absorption spectrophotometer and comparing the response obtained with those of the standard solutions.

In developing the procedure, the following observations were made:

1. Sulphuric and nitric acids in the presence of iron depress the chromium absorption. The presence of hydrochloric acid in slight excess does not give rise to this effect.
2. Neutralization with 0·88 ammonia, prior to reacidification with hydrochloric acid, gives rise to erratic results. The reason for this is not fully understood. Potassium hydroxide, which does not give rise to this effect, is therefore used.
3. Extensive preliminary work showed that it was necessary to add lanthanum chloride, but not iron, to the standards. Lanthanum chloride not only overcomes all depressive interference from iron, but also gives rise to a slight enhancement.

Method II. With the nitrous oxide–acetylene flame, iron does not interfere with the absorption of chromium, although excess acidity in the solutions does still exert an effect. If the older type argon-filled lamp is used, the flame background of this gas system gives rise to an unacceptably noisy signal at the more sensitive 3594 Å and 3579 Å wavelengths, so that it is necessary to use the line at 4254 Å. At this wavelength the optimum concentration range for determinations in aqueous solutions using the nitrous oxide–acetylene flame is from 25–250 μg/ml. With the newer high-spectral-output lamps that are neon-filled the more sensitive wavelengths can be used and the optimum range for determination is from 5 to 30 μg/ml.

The sample, chosen so that it will yield a test solution containing chromium in the optimum range for determination, is weighed and dissolved as described in Method I. Excess acid is neutralized with potassium hydroxide, and the solution just reacidified with hydrochloric acid. Recent work has indicated that the most reliable results for this important determination are obtained by using the nitrous oxide–acetylene flame and employing B.C.S. steels of similar composition to the material under test to prepare the standard solutions.

References will be found in the literature to procedures for this determination that utilize the air–acetylene flame and that overcome the depressive interference of iron by the addition of a very heavy excess of ammonium chloride (1 to 2 per cent in the final test solution[(56,59)]). These procedures do, we find, give rise to accurate results in some cases, but necessitate also the use of standards that contain iron at about the same level as the test solutions. Because the suppressant (NH_4Cl) has to be present at such a high concentration level, and because the use of matching standards is still necessary, it may be felt that the procedures described earlier are more convenient to the analyst.

Analyses of Silicates, including Glass, Ceramics, Coal Ash, Minerals etc.

Atomic absorption spectroscopy provides a reliable and rapid method for the determination of the metallic constituents of silicate minerals. Our own investigations (E.E.L.) have centred upon ceramic materials in which the elements of main interest are aluminium, iron, calcium, magnesium, sodium, and potassium, but the following method is generally applicable to most silicates. It consists essentially of digesting the mineral with hydrofluoric–perchloric–nitric acid mixture, so that silicon is volatilized as the tetrafluoride. The residue is dissolved in hydrochloric acid, and lanthanum chloride added to overcome suppression of the calcium response by aluminium. This solution is adjusted to volume and used for determinations.

Procedure

This is best illustrated by considering the analysis of a typical earthenware body, which is made up as follows:

SiO_2	71·25%		
Al_2O_3	18·00%	Al	9·52%
K_2O	1·50%	K	1·25%
CaO	0·80%	Ca	0·57%
Na_2O	0·75%	Na	0·56%
Fe_2O_3	0·50%	Fe	0·35%
MgO	0·30%	Mg	0·18%
TiO_2	0·25%		
Loss of volatiles	6·65%		

If 0·5 g of such a body is attacked and the contents eventually placed in 500 ml of solution, that solution will contain:

Al	95·2 μg/ml
K	12·5 μg/ml
Ca	5·7 μg/ml
Na	5·6 μg/ml
Fe	3·5 μg/ml
Mg	1·8 μg/ml

Sample Preparation. 0·5 g of the silicate is weighed into a platinum crucible and treated with 10 ml of hydrofluoric acid (A.R.), 5 ml of 1 + 1 nitric acid + water, and 5 ml of 1 + 4 perchloric acid + water. The digest is evaporated almost to dryness on a sand bath, the residue treated with a further 5 ml of 1 + 4 perchloric acid + water, and again evaporated to near dryness. About ten drops of concentrated hydrochloric acid are added to the crucible, followed by 15 ml of water. The mixture is then digested on a steam bath for 20 minutes, cooled, and transferred to a 500 ml graduated flask. 10 ml of a lanthanum chloride solution containing 65000 μg/ml La are added and the volume made up to the mark with deionized water.

Standard Solutions. Complex standard solutions are prepared. For the earthenware body described above convenient standards would have the following compositions:

	Standard solution I (μg/ml)	Standard solution II (μg/ml)	Standard solution III (μg/ml)
La	6500	6500	6500
Al	50	100	150
K	5	10	15
Ca	2·5	5	7·5
Na	2·5	5	7·5
Fe	2	5	10
Mg	0·5	1	2

Determination. The determinations of all the elements are straightforward, and the analyst is referred to Chapter IV for specific details of their characteristics. For this application it is preferable to use the nitrous oxide-acetylene flame for the estimations of calcium and magnesium. The spectrophotometer would need to be fitted with a suitable red sensitive photomultiplier for the potassium determination.

When it is required to determine silica, as well as the metallic constituents, by atomic absorption the most useful procedure for sample attack is with lithium borate. A 0·2 g sample requires 1 g of $LiBO_2$. The fusion is placed while still molten into 100 ml of 3 per cent nitric acid. Standards used for comparison should be prepared so as to contain about the same loading of lithium borate as the test solutions.

Analysis of Cement

To a first approximation, the composition of a typical cement is as follows:

CaO	65%	Ca	46·45%
SiO_2	21%		
Al_2O_3	6%	Al	3·2%
Fe_2O_3	3%	Fe	2·1%
MgO	1%	Mg	0·60%
SO_3	1%		
K_2O	1%	K	0·83%
Na_2O	0·25%	Na	0·19%
TiO_2	0·25%		
P_2O_5	0·25%		
MnO	0·10%	Mn	0·08%
Loss on ignition	1·15%		

Of these constituents aluminium, iron, calcium, magnesium, sodium, potassium, and manganese can be determined in solution by atomic absorption spectroscopy, without separation from other constituents or from one another. Silica can also be determined, but in order to do so it is necessary to get the element into solution. It has been suggested that the lithium borate fusion, outlined in the section on silicate analysis, should prove suitable for this step.

In an alternative procedure[81] that has been described for the determination of silicon in cements, a 0·4 g sample is weighed into a polythene beaker, wetted with 10 ml of water and attacked with 6 ml of concentrated hydrochloric acid and 0·5 ml of 40% hydrofluoric acid.

Twenty ml of a solution containing 2·5% vanadium and 10 ml of a 5000 μg/ml Na solution are added to the digest, which is then made up to 100 ml in a calibrated polythene cylinder. Measurement is made against standards comparable with the test solution. The other constituents of the cement can also be determined by dilution of the water analyte.

Principle of a Method in which SiO_2 is not determined

For a cement approximating to the above composition, 1 g of cement is dissolved in hydrochloric acid and the solution filtered from the insoluble silica. The solution is adjusted to a volume of 200 ml with deionized water, and will then contain:

Ca	2322·5 μg/ml
Al	160 μg/ml
Fe	105 μg/ml
K	41·5 μg/ml
Mg	30 μg/ml
Na	9·5 μg/ml
Mn	4 μg/ml

Complex standard solutions are prepared to resemble the cement extract, and to cover the anticipated range of concentrations for the elements.

Aluminium and manganese can then be determined by spraying the solution prepared from the cement against these standards.

By diluting the complex standards and test solution(s) ten times, and adding lanthanum to suppress possible interference from aluminium on the magnesium response, the estimations of iron, magnesium, sodium, and potassium may be effected.

The determination of calcium, to the required degree of precision, is the most difficult feature of this analysis. Experimental work established that the optimum concentration range for the determination of calcium was, for this application, between about 4 and 6 μg/ml in solution. In order to achieve an accuracy of ±1% on the CaO content (i.e. ±0·7% on the calcium content of the cement) it is necessary to be able to discern differences better than 0·07 μg/ml calcium in the test solutions. This requirement is achieved by employing the procedure described in Chapter III in the section on the determination of metals present as macro constituents of a sample.

Procedure for an Analysis in which Estimation of SiO_2 is *not* required

Sample Preparation (taken from C.E.R.I.L.H. technical publication No. 172, p. 32)

Reagents hydrochloric acid (A.R.) 36%
gelatine solution 2·5%

Weigh 1 g of cement into a 250 ml tall-form beaker. Add 15 ml of concentrated hydrochloric acid and, when the initial reaction has subsided, boil gently for about six minutes. Cool to 70°C and add 5 ml of gelatine (2·5% solution) also at a temperature of 70°C. Filter the flocculated silica from the clear liquor, wash the precipitate with hot water, cool, transfer the filtrate and washings to a 200 ml graduated flask and dilute to the mark. This solution is the master extract.

Standard Solutions. Prepare complex standard solutions. For the cement described above convenient standards would have the following compositions:

Standard solution I (μg/ml)		Standard solution II (μg/ml)		Standard solution III (μg/ml)	
Ca	2,000	Ca	2,500	Ca	3,000
Al	100	Al	150	Al	200
Fe	50	Fe	100	Fe	200
K	10	K	20	K	50
Mg	10	Mg	20	Mg	50
Na	5	Na	10	Na	20
Mn	1	Mn	2	Mn	5

The determination of the elements other than calcium is straightforward. Aluminium and manganese would be determined on the sample solution, prepared as described above, by comparison with the master standard. Magnesium, iron, sodium, and potassium would be determined upon aliquots prepared by transferring 20 ml of the master extract and above standards, to 200 ml graduated flasks, adding to each flask 20 ml of stock lanthanum chloride solution (65000 ppm lanthanum) and diluting to the mark with deioinzed water.

For this application the magnesium determination is best carried out with the nitrous oxide–acetylene flame system.

Preparation of Test and Standard Solutions for Calcium Determination. Re-dilute the diluted extracts prepared for the determinations of iron etc. fifty times to yield solutions containing from 4–6 ppm calcium.

Pipette 5 ml of each diluted extract and diluted standard to a separate 200 ml graduated flask. Add 20 ml of lanthanum chloride stock solution (65000 ppm lanthanum) and dilute to the mark with deionized water.

The atomic absorption determination is then carried out with the nitrous oxide–acetylene flame, utilizing scale expansion in conjunction with suppression of the zero, so as to set the 4 μg/100 ml standard to zero on the expansion unit and the 6 μg/100 ml standard to full scale, as described in Chapter III (method 7).

It is advantageous if the expanded signal can be fed to a chart recorder. Under normal routine conditions an accuracy of about 0·2 per cent upon the Ca content of the cement is attainable.

Soil Analysis

This was one of the first areas in which atomic absorption spectroscopy found application. The agricultural chemist is required to estimate a surprisingly large number of metals, of which calcium, magnesium, potassium, sodium, lithium, manganese, cobalt, nickel, copper, zinc, iron, lead, strontium, cadmium, chromium, aluminium, barium, mercury, and molybdenum can be determined reliably and quickly in trace quantities by atomic absorption. The freedom of the technique from spectral interferences allows metals to be determined in extract solutions and run-off waters without prior chemical separations; indeed at the low levels present in these solutions even chemical interferences are usually absent.

Soil analysis can be divided into two main sections (a) total elemental analysis and (b) cation exchange analysis (the leachable metal contents of the soil).

Total Elemental Analysis

The method is described in more detail under silicate analysis and consists essentially of digesting the dried or ignited soil with hydrofluoric–perchloric–nitric acid mixture, so that silicon is volatilised as the tetrafluoride. The residue is dissolved in hydrochloric acid, lanthanum chloride added to overcome suppression of the calcium response by aluminium, and adjusted accurately to a suitable volume.

Complex standard solutions are prepared to cover the expected concentration range of the constituents, and the determinations are made by aspirating standards and test solutions to the spectrophotometer and comparing the responses. The determinations of all the elements likely to be required are straightforward and the reader is referred to Chapter IV for further detailed information that may be required.

For this application, calcium and magnesium are most reliably determined with the nitrous oxide–acetylene flame system. Potassium determined with the air–acetylene flame will require the use of a red-sensitive photomultiplier.

Metals that are present in only very low concentrations will require the use of scale expansion.

If manganese is to be determined in solutions that contain perchloric acid it is essential to reduce the metal to the divalent state by the action of sodium nitrite or hydroxylamine hydrochloride. See the section on manganese in Chapter IV.

Cation Exchange Analysis

The cation exchange determinations that can be advantageously performed by atomic absorption spectroscopy are:

1. Measurement of the exchangeable metallic cations.
2. Measurement of cation exhange capacity of the soil.

These measurements then allow the chemist to determine the percentage exchange saturation of the soil and the percentage of saturation with alkali cations.

Exchangeable Metallic Cations are eluted from the soil either by continuous or repeated washing of a sample. Continuous washing requires the use of a percolation device such as a funnel or porous porcelain crucible. Repeated washings may be accomplished by repeatedly dispersing the sample in fresh portions of solvent and removing the eluate by centrifuging. Full practical details of the procedures employed and precautions to be taken are to be found in *Soil Chemical Analysis* by M. L. Jackson (University of Wisconsin, 1965).

The exchangeable metallic cations most frequently determined are calcium, magnesium, potassium, manganese, and sodium. They may all be estimated in an extract obtained by eluting the soil with 1 N ammonium acetate solution. This solvent possesses the advantages that it effectively wets a soil, efficiently replaces the cations, and is suitable for flame photometric determinations.

The less frequently determined exchange metals (aluminium, iron, copper, zinc, and cobalt) may also be estimated in the ammonium acetate extract, but are more reliably determined when extracted by alternative procedures, e.g. copper is best extracted with a solution of E.D.T.A. disodium salt.

In the procedure employing ammonium acetate, the soil sample, preferably in the field-moist condition, is passed through a 2 mm sieve and 50–100 g weighed into a 250 ml flask. 100 ml of 1 N ammonium acetate solution are added, the flask stoppered, shaken, and the contents allowed to stand overnight. The contents are transferred to a Buchner funnel fitted with a Whatman No. 42 filter paper. After the filtrate has been separated from the solid, washing is carried out, over a period of about 1 hour, with a further 300 ml of ammonium acetate solution.

The extract is transferred to a 500 ml graduated flask, treated with 50 ml of lanthanum chloride solution (65000 μg/ml La) and made up to the mark with 1 N ammonium acetate solution (the lanthanum chloride addition is made to overcome possible phosphate interference). This test solution is used for the determinations of all the commonly sought metals. Complex standards, prepared so as to resemble approximately the composition of the test solution and to cover the estimated concentration ranges of the metals to be determined, are used.

The results are normally related to the soil in 'the 100°C oven-dried condition' by calculation.

Cation Exchange Capacity. The eluted soil from the Buchner funnel is conveniently used for this determination, which involves the measurement of the total quantity of negative charges per unit weight of the soil. This capacity is measured by reloading the ammonium acetate leached soil with a 1 N solution of calcium chloride, buffered to a neutral or slightly alkaline *p*H, and then washing out the excess salt with an electrolyte-free solvent, such as 80 per cent acetone. The calcium is now replaced by further washing with 1 N ammonium acetate and is determined in the eluate as described previously to give the total exchange capacity of the soil.

An extremely large number of methods for estimating the cation exchange capacity of soils is provided by different combinations of pretreatment cation salt, solvent, and washing procedure. Atomic absorption spectroscopy provides a reliable and rapid method for the final estimation in every case.

Trace Metals in Water and Effluents

The mineral contents of natural waters, feed waters, and drinking supplies vary widely in both the nature and concentration of the metals present in them. This metal content can affect the potability, and suitability of a water as a source for public supply. For example, the following elements, if present in more than the stated concentrations (United States figures) render a water unpotable:

Cu 1·5 mg/l
Zn 1·5 mg/l
Mn 5 mg/l
Fe 50 mg/l
Magnesium + sodium sulphate 1000mg/l

The toxic metals listed below, if present in more than the stated quantities (United States figures) preclude a natural water from use as a source of public supply.

Cd 0·01 mg/l
Cr 0·05 mg/l
Pb 0·05 mg/l

Again, water to be used as a feed to modern high pressure boilers must contain less than 5 μg per litre of copper or iron.

Effluent waters and sewage can contain an even larger variety of metals covering a much wider range of concentrations. The measurement and control of these pollutents constitutes an important part of the activities of river authorities, drainage boards, manufacturing companies, and health authorities. Atomic absorption spectroscopy offers the most versatile, reliable, simple, and rapid procedure available to the chemist for many of these important analyses.

Interferences

Even with the most heavily polluted effluents it is unlikely that any of the metals will be present in so heavy a concentration as to exert an interfering effect upon the absorption of others present—a very significant advantage to the analyst. It is nevertheless convenient to use combined standards, each containing all the metals that are to be analysed, in ranges that cover the anticipated concentrations in the effluent, because this procedure will reduce the number of flasks required. Any small inter-element effects will, of course, be cancelled out by the adoption of this procedure.

Reference to the relevant section of Chapter IV will give a good indication of the few serious interferences likely to be encountered. For convenience they are listed below:

(i) Depression caused by the presence of phosphate upon the absorption of the alkaline earths.
(ii) Depression caused by the presence of aluminium upon the absorption of the alkaline earths.

(iii) Bicarbonate interference upon calcium absorption.
(iv) Depression caused by the presence of iron upon the absorption of chromium.

All four of these interferences are overcome by the addition of lanthanum chloride to the test and standard solutions.

Types of Determination

Analyses for metallic constituents of waters and effluents may conveniently be divided into three categories.

(i) Those in which the element to be determined is present in sufficient quantity for a reading to be made directly without concentration or scale expansion (Ca, Mg, Na, K).
(ii) Those in which the element is present in solution in only sufficient quantity for a reading to be made directly upon the sample by scale expanding the signal (Fe).
(iii) Those in which the metal is present in so low a concentration that a prior extraction into an organic solvent is necessary (Mo, Al, Cu, Zn).

The concentration ranges over which metals can be directly determined and the limits of detection in aqueous solution are given in Chapter IV under the characteristics of each element.

Concentration Procedures

The most valuable and commonly used procedure for extractive concentration of heavy metals is that employing chelation with ammonium pyrrolidine dithiocarbamate (A.P.D.C.) followed by extraction into isobutyl methyl ketone (I.B.M.K.) This is described in Chapter III.

Application to Petroleum Analysis

One of the major advantages that atomic absorption spectroscopy offers to the analyst is that solutions need not be confined to aqueous media. Time-consuming digestions and extractions necessitating the running of reagent blanks and prone to handling losses, are thus avoided. This advantage is of particular value to the petroleum chemist, as it enables direct determinations to be made of aluminium and vanadium in crude and fuel oils, and of additives such as zinc, calcium and barium in lubricating oils.

PROCEDURE

Oils and greases must be diluted with a suitable solvent before they can be aspirated to the atomic absorption spectrophotometer. References will be found in the literature that recommend the use of hydrocarbons such as iso-octane, *n*-heptane and xylene for this dilution. Such solvents alone are not always suitable for use with every instrument since they may give rise to such a rich smoky flame that it may become necessary (in order to establish non-luminous conditions) to restrict the acetylene flow to the point where the flame eventually lifts off the burner. Mixtures of white spirit and acetone, white spirit and isopropyl alcohol and iso-octane and isobutyl methyl ketone overcome this difficulty.

Cyclohexanone is also suitable and with these diluents solutions may be prepared that allow determinations to be carried out safely with both the air–acetylene and nitrous oxide–acetylene flames.

The standards for such determinations must be prepared from suitable oil-soluble compounds, or from previously analysed oils so as to resemble the test solutions.

In general, due to improved solution uptake, determinations carried out in non-aqueous media will be more sensitive than those carried out in aqueous solutions.

The wear metals iron, copper, silver, magnesium, chromium, tin, lead, and nickel may be determined in lubricating oils. The oils are diluted with a suitable solvent prior to analysis. Standard solutions prepared by dissolving oil-soluble compounds, usually naphthenates, in a virgin oil, and then diluting the solution obtained with the relevant solvent, are used for comparison.

The estimation is only approximate since the wear metals will be present as particles suspended in the oil, and some of these particles may be too large to be converted into an atomic vapour in the optical path through the flame.

For more accurate determinations of wear metals it is necessary to ash the oil sample, dissolve the residue with acid, and carry out the measurement in an aqueous medium. Even when this more prolonged sample preparation is resorted to, atomic absorption still offers the most reliable procedure for final measurement of metal concentrations.

INTERFERENCES

Of the metals normally determined in oils, very few are liable to chemical interferences. The presence of aluminium depresses the absorption of calcium, but the estimation of this metal is normally required only in refined oils where aluminium is either absent or present only in very low quantity. The nitrous oxide–acetylene flame system can be used to overcome this interference, but with this system it is advisable always to make an addition of an oil-soluble sodium or potassium compound to the test and standard solutions. It is of interest to note the presence of phosphate in a non-ionic medium does not depress the absorption of calcium.

The absorption of chromium in the air–acetylene flame is depressed in the presence of iron, but this depression can again be overcome by using the nitrous oxide–acetylene system.

The two elements whose estimations are most commonly required in crude and fuel oils are aluminium and vanadium; they require the use of the nitrous oxide–acetylene flame and are free from interferences.

Determination of Lead in Petrol

This is a determination that possesses peculiar characteristics, and for that reason it is discussed separately. The two compounds most commonly employed as anti-knock additives are tetraethyl and tetramethyl lead, but in order to apply the method directly it is necessary to know which of these, or what mixture of the two, has been included in the sample to be analysed. If the analyst possesses no information as to the nature of the lead additive, direct determination is unreliable. In such cases destruction of the lead alkyl followed by extraction of the lead into an aqueous phase is the most satisfactory procedure.

(A) INDIRECT DETERMINATION (after Shell United Kingdom Ltd)

Light petroleum distillates can contain lead at the μg/ml level, and the determination of this element is commonly effected by polarography, after extraction into an aqueous phase. This procedure requires about $2\frac{1}{2}$ hours per determination. The following procedure, developed by Shell United Kingdom Ltd. and employing atomic absorption, requires only about 45 minutes.

The sample is treated with bromine to convert the lead to lead bromide which is extracted with dilute nitric acid, and the lead content of the extract determined on the atomic absorption spectrophotometer, if necessary fitted for scale expansion. The following standard solutions and reagents are required.

Master Stock Solution (5000 *μg/ml*). Dissolve 3·995 g of lead nitrate, $Pb(NO_3)_2$(A.R.), in approximately 200 ml of deionized water, transfer to a 500 ml graduated flask, and dilute to the mark with deionized water.

Intermediate Stock Solution (100 *μg/ml*). Pipette 10 ml of the master solution to a 500 ml graduated flask, and dilute to the mark with deionized water.

Working Standards. Prepare working standards containing 50, 25, 10 and 5 μg/ml lead by dilution from the intermediate solution.

Blank Solution. 1 per cent nitric acid (8 parts by volume nitric acid s.g. 1·42 (A.R.) + 992 parts deionized water).

Digestion Reagent. Bromine (A.R.).

Sample Preparation

1. 250 ml of sample are measured into a 600 ml beaker and bromine added dropwise until a deep wine colour persists for 2 minutes. The sample is then allowed to stand for a further 10 minutes to ensure complete bromination.
2. The solution is evaporated on a hotplate until the volume is reduced to approximately 50 ml.
3. The contents of the beaker are transferred to a 100 ml separating funnel and the beaker washed with a 10 ml portion of 1 per cent nitric acid which is then added to the separating funnel.
4. The separating funnel is shaken vigorously for 2 minutes, the two phases allowed to settle and the lower layer run into a 60 ml beaker.
5. The extraction is repeated with a second 10 ml portion of 1 per cent nitric acid, and the extracts combined and made up to 25 ml in a graduated flask with 1 per cent nitric acid.
6. This acid extract is now ready for analysis by atomic absorption spectroscopy.

Measurement. The concentration of lead in the 25 ml extracts obtained from the samples is at most about 50 μg/ml. Determination of lead at this concentration in aqueous solution is now possible on most atomic absorption spectrophotometers without scale expansion. The procedure was developed in 1966 when lead lamps were greatly inferior to the ones now available. For this reason the authors used scale expansion.

It is recommended that 1 per cent nitric acid should be aspirated to the spectrophotometer after each lead-containing solution, so as to clear the system.

Calculation of Results and Discussion. An exhaustive series of investigations carried out by Shell United Kingdom Ltd, indicated that both the procedure and instrument readings were highly repeatable, and that over the narrow concentration range experienced with control samples the response was linear, so that the following calculation was valid.

Calculation of Results. The standard is selected having the nearest scale reading to that of the sample, and the lead content of the original sample of distillate calculated from:

$$\text{Pb as } \mu\text{g/ml } (wt/wt) = \frac{S_0 - S_1}{S_0 - S_2} \times C \times \frac{1}{10 \times \text{s.g. of original sample}}$$

where

S_0 = scale reading for blank, i.e. 1 per cent ntiric acid
S_1 = scale reading for sample extract
S_2 = scale reading for standard
C = concentration of standard Pb μg/ml (wt/wt)

(B) Direct Determination

If the nature of the anti-knock additive is known it is possible to determine lead directly in petrol by atomic absorption.

Response curves for solutions of tetraethyl lead, tetramethyl lead and mixtures of these two compounds obtained at 2833 Å using air–acetylene, are shown in Fig. 5.3.

As stated previously, besides pure T.E.L. and pure T.M.L., mixtures of these compounds are used, which will of course exhibit different response characteristics. These mixtures, though, are almost invariably prepared so that they have the same lead content as either the straight T.E.L. or T.M.L. compounds (112.25 g Pb per litre).

Since the master additives have identical lead contents it is possible for blenders to utilize atomic absorption to check that these master blends have been properly constituted; for example, response curves as shown in Fig. 5.3 could be constructed from dilutions of T.E.L. or T.M.L. mixtures whose proportions are known. Works batches could then be compared against these curves and prediction as to their relative contents of T.M.L. and T.E.L. made.

An investigation has been carried out in our laboratories (E.E.L.) to determine the effect of different hydrocarbons on the lead absorption. Identical response curves were obtained for T.E.L. originally diluted with xylene, iso-octane and cyclohexane.

The effects of the presence of differing bromo and chloro compounds has also been investigated and found not to affect the absorption.

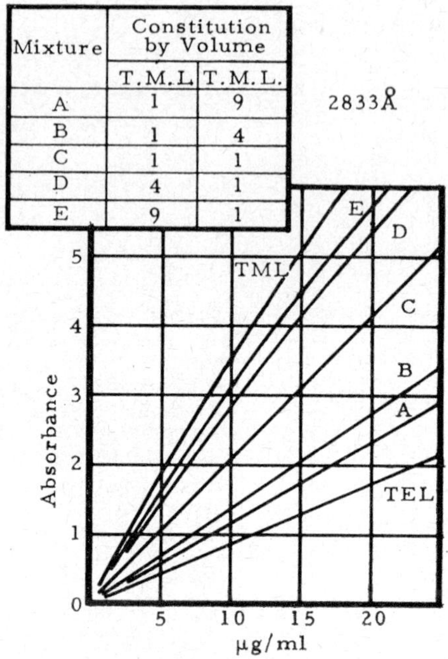

Fig. 5.3 Typical response curves; TEL/TML

Preparation of Standard Solutions. Tetraethyl lead and tetramethyl lead are supplied by the Associated Octel Co. Ltd, in diluted form under the designations 'Dilute T.E.L.-B compound' and 'Dilute T.M.L.-CB compound'. Both contain 112·25 g lead/litre, in a medium of 70 per cent v/v xylene and 30 per cent v/v *n*-heptane. These compounds are highly toxic and Associated Octel's recommended safety precautions must be rigidly adhered to.

Prepare an intermediate standard (containing 560 μg/ml) by diluting 1 ml of either T.E.L.-B or T.M.L.-CB (according to which has been included in the sample to be analysed) to 200 ml with 'raw-petrol'.

Working Standards. Working standards are prepared from the intermediate standard solution by dilution with cyclohexanone or white spirit and isopropyl alcohol. It is essential to use an oxygenated compound for this operation, as further dilution with a hydrocarbon gives rise to a medium that produces a luminous flame at all pressures of acetylene and air obtainable on the instrument.

The standards should cover the range 0–15 μg/ml lead if the tetramethyl compound is being estimated, and 0–50 μg/ml if the tetraethyl compound is to be assessed.

Blank Solution. Use the relevant diluent.

Sample Preparation. Dilute the petrol sample with cyclohexanone or white spirit and isopropyl alcohol, so that its estimated lead content lies in the relevant range stated above.

Measurement. Aspirate this sample and the standards to the atomic absorption spectrophotometer and by comparing the readings obtain the lead content of the petrol.

Chap. V References

BIOLOGICAL MATERIAL

1. WILLIS, J. B., The determination of metals in blood serum by atomic absorption spectroscopy I, calcium. *Spectrochim. Acta* 16 (1960) 259.
2. WILLIS, J. B., The determination of metals in blood serum by atomic absorption spectroscopy II, magnesium. *Spectrochim. Acta* 16 (1960) 273.
3. WILLIS, J. B., The determination of metals in blood serum by atomic absorption spectroscopy III, sodium and potassium. *Spectrochim. Acta* 16 (1960) 551.
4. WILLIS, J. B., The determination of calcium and magnesium in urine by atomic absorption spectroscopy. *Anal. Chem.* 33 (1961) 556.
5. DAWSON, J. B., and HEATON, F. W., The determination of magnesium in biological materials by atomic absorption spectrophotometry. *Biochem. J.* 80 (1961) 99.
6. NEWBURN, E., Application of atomic absorption spectroscopy to the determination of calcium in saliva. *Nature* 192 (1961) 1182.
7. DAVID, D. J., Atomic absorption spectrochemical analyses of plant materials with particular reference to manganese and iron. *At. Abs. Newsletter* 1 (1962) 45.
8. WILLIS, J. B., Determination of lead and other heavy metals in urine by atomic absorption spectroscopy. *Anal. Chem.* 34 (1962) 614.
9. PARKER, H. E., Magnesium, calcium and zinc in animal nutrition. *At. Abs. Newsletter* 2 (1963) 23.

10. Berman, E., The determination of lead in blood and urine by atomic absorption spectrophotometry. *At. Abs. Newsletter* 3 (1964) 111.

11. Buchanon, J. R., and Muraoka, T. T., Determination of zinc and manganese in tree leaves by atomic absorption spectroscopy. *At. Abs. Newsletter* 3 (1964) 79.

12. Frey, S. W., The determination of copper, iron, calcium, sodium and potassium in beer by atomic absorption spectrophotometry. *At. Abs. Newsletter* 3 (1964) 127.

13. Fuwa, K., Pulido, P., McKay, R., and Vallee, B. L., Determination of zinc in biological materials by atomic absorption spectrophotometry. *Anal. Chem.* 36 (1964) 2407.

14. Morgan, M. E., Determination of copper in milk by atomic absorption spectroscopy. *At. Abs. Newsletter* 3 (1964) 43.

15. Slavin, W., Sprague, S., Pieders, F., and Cordova, V., The determination of certain toxicological trace metals by atomic absorption spectrophotometry. *At. Abs. Newsletter* 4 (1964) 7.

16. Trent, D. J., and Slavin, W., Factors in the determination of strontium by atomic absorption spectrophotometry with particular reference to ashed biological samples. *At. Abs. Newsletter* 3 (1964) 53.

17. Wacker, W. E. C., Iida, C., and Fuwa, K., Accuracy of determinations of serum magnesium by flame emission and atomic absorption spectrophotometry. *Nature* 202 (1964) 659.

18. Slavin, W., and Sprague, S., The determination of trace metals in blood and urine by atomic absorption spectrophotometry. *At. Abs. Newsletter* 4 (1964) 1.

19. Sprague, S., and Slavin, W., The determination of nickel in urine by atomic absorption spectrophotometry—preliminary study. *At. Abs. Newsletter* 3 (1964) 160.

20. Berman, E., Application of atomic absorption spectrometry to the determination of copper in serum, urine and tissue. *At. Abs. Newsletter* 4 (1965) 296.

21. Slavin, W., Applications of atomic absorption spectroscopy in the food industry. *At. Abs. Newsletter* 4 (1965) 330.

22. Sprague, S., and Slavin, W., Determination of iron, copper and zinc in blood serum by an atomic absorption method requiring only dilution. *At. Abs. Newsletter* 4 (1965) 228.

23. SLAVIN, W., The determination of various metals in synthetic fibres using atomic absorption spectrophotometry. *At. Abs. Newsletter* 4 (1965) 192.

24. GUILLAUMIN, R., Determination of calcium and magnesium in vegetable oils and fats by atomic absorption spectrophotometry. *At. Abs. Newsletter* 5 (1966) 19.

25. PRÉVÔT, A., Determination of sodium and potassium in oils and fats. *At. Abs. Newsletter* 5 (1966) 13.

26. CHANG, T. L., GOVER, T. A., and HARRISON, W. W., Determination of magnesium and zinc in human brain tissue by atomic absorption spectroscopy. *Anal. Chim. Acta* 34 (1966) 17.

27. KAHNKE, M. J., Atomic absorption spectrophotometry applied to the determination of zinc in formalinized human tissue. *At. Abs. Newsletter* 5 (1966) 7.

28. SPRAGUE, S., and SLAVIN, W., A simple method for the determination of lead in blood. *At. Abs. Newsletter* 5 (1966) 9.

29. BERMAN, E., Determination of cadmium, thallium and mercury in biological materials by atomic absorption. *At. Abs. Newsletter* 6 (1967) 57.

30. Metallic impurities in organic matter. Sub-committee, *Analyst* 92 (1967) 403. The use of 50 per cent hydrogen peroxide for the destruction of organic matter.

31. ROACH, A. G., SANDERSON, P., and WILLIAMS, D. R., Determination of trace amounts of copper, zinc and magnesium in animal feeds by atomic absorption spectrophotometry. *Analyst* 93 (1967) 42.

32. HARTLEY, F. R., and INGLIS, A. S., The determination of aluminium in wool by atomic absorption spectroscopy. *Analyst* 92 (1967) 622.

33. PIPER, K. G., and HIGGINS, G., Estimation of trace metals in biological material by atomic absorption spectrophotometry. *Proc. Assn. Clin. Biochemists* IV (1967) 190.

34. HARTLEY, F. R., and INGLIS, A. S., The determination of metals in wool by atomic-absorption spectroscopy. *Analyst* 93 (1968) 394

METALLURGICAL

35. RAWLING, B. S., GREAVES, M. C., and AMOS, M. D., The determination of silver in lead concentrates by atomic absorption spectroscopy. *Nature* 188 (1960) 137.

36. GIDLEY, J. A. F., and JONES, J. T., The determination of zinc in metallurgical materials by atomic absorption spectrophotometry. *Analyst* 85 (1960) 249; 86 (1961) 271.

37. ELWELL, W. T., and GIDLEY, J. A. F., The determination of lead in copper-base alloys and steel by atomic absorption spectrophotometry. *Anal. Chim. Acta* 24 (1961) 71.

38. ANDREW, T. R., and NICHOLS, P. N. R., The application of atomic absorption to the rapid determination of magnesium in electronic nickel and nickel alloys. *Analyst* 87 (1962) 25.

39. BELCHER, C. B., and BRAY, H. M., The determination of magnesium in iron by atomic absorption spectrophotometry. *Anal. Chim. Acta* 26 (1962) 322.

40. MCPHERSON, G. L., PRICE, J. W., and SCAIFE, P. H., Application of atomic absorption spectroscopy to the determination of cobalt in steel, alloy steel and nickel. *Nature* 199 (1963) 371.

41. KINSON, K., HODGES, R. J., and BELCHER, C. B., The determination of chromium in low-alloy irons and steels by atomic absorption spectrophotometry. *Anal. Chim. Acta* 29 (1962) 134.

42. BELCHER, C. B., and KINSON, K., The determination of manganese in iron and steel by atomic absorption spectrophotometry. *Anal. Chim. Acta* 30 (1964) 483.

43. KINSON, K., and BELCHER, C. B., The determination of nickel in iron and steel by atomic absorption spectrophotometry. *Anal. Chim. Acta* 30 (1964) 64.

44. KINSON, K., and BELCHER, C. B., Determination of minor amounts of copper in iron and steel by atomic absorption spectrophotometry. *Anal. Chim. Acta* 31 (1964) 180.

45. SHAFTO, R. G., The determination of copper, iron, lead and zinc in nickel plating solutions by atomic absorption. *At. Abs. Newsletter* 3 (1964) 115.

46. SPRAGUE, S., and SLAVIN, W., The determination of copper, nickel, cobalt, manganese and magnesium in iron and steels by atomic absorption spectrophotometry. *At. Abs. Newsletter* 3 (1964) 72.

47. WILSON, L., The determination of silver in aluminium alloys by atomic absorption spectroscopy. *Anal. Chim. Acta* 30 (1964) 377.

48. DYCK, R., The determination of chromium, magnesium and manganese in nickel alloys by atomic absorption spectrophotometry. *At. Abs. Newsletter* 4 (1965) 170.

49. McPherson, G. L., Atomic Absorption spectrophotometry as an analytical tool in a metallurgical laboratory. *At. Abs. Newsletter* 4 (1965) 186.

50. Beyer, M., Determination of manganese, copper, chromium, nickel and magnesium in cast iron and steel. *At. Abs. Newsletter* 4 (1965) 212.

51. Farrar, B., Determination of copper and zinc in ore samples and lead-base alloys. *At. Abs. Newsletter* 4 (1965) 325.

52. Humphrey, J. R., Determination of magnesium in uranium by atomic absorption. *Anal. Chem.* 37 (1965) 1604.

53. Bell, G. F., The analysis of aluminium alloys by means of atomic absorption spectrophotometry. *At. Abs. Newsletter* 5 (1966) 73.

54. Capacho-Delgado, L., and Manning, D. C., Determination of vanadium in steels and gas oils. *At. Abs. Newsletter* 5 (1966) 1.

55. Kirkbright, G. F., Smith, A. M., and West, T. S., Rapid determination of molybdenum in alloy steels by atomic absorption spectroscopy in a nitrous oxide–acetylene flame. *Analyst* 91 (1966) 700.

56. Mostyn, R. A., and Cunningham, A. F., Determination of molybdenum in ferrous alloys by atomic absorption spectrometry. *Anal. Chem.* 38 (1966) 121.

57. Mansell, R. E., Emmel, H. W., and McLaughlin, E. L., Analysis of magnesium and aluminium alloys by atomic absorption spectroscopy. *Appl. Spectroscopy* 20 (1966) 231.

58. Kirkbright, G. F., Peters, M. K., and West, T. S., Determination of small amounts of molybdenum in niobium and tantalum by atomic absorption spectroscopy in a nitrous oxide–acetylene flame. *Analyst* 91 (1966) 705.

59. Barnes, L., Jr., Determination of chromium in low alloy steels by atomic absorption spectrometry. *Anal. Chem.* 38 (1966) 1083.

60. Dagnall, R. M., West, T. S., and Young, P., Determination of trace amounts of lead in steels, brass and bronze alloys by atomic absorption spectrometry. *Anal. Chem.* 38 (1966) 358.

61. Kirkbright, G. F., Peters, M. K., and West, T. S., Determination of trace amounts of copper in niobium and tantalum by atomic absorption spectroscopy. *Analyst* 91 (1966) 411.

62. Wilson, L., The determination of cadmium in stainless steel by atomic absorption spectroscopy. *Anal. Chim. Acta* 35 (1966) 123.

63. BELL, G. F., On the effect of copper on the determination of zinc in aluminium. *At. Abs. Newsletter* 6 (1967) 18.

64. MCAULIFFE, J. J., A method for determination of silicon in cast iron and steel by atomic absorption spectrometry. *At. Abs. Newsletter* 6 (1967) 69.

65. JURSIK, M. L., Application of atomic absorption to the determination of Al, Fe, and Ni in the same uranium-base sample. *At. Abs. Newsletter* 6 (1967) 21.

66. SCOTT, T. C., ROBERTS, E. D., and CAIN, D. A., Determination of minor constituents in ferrous materials by atomic absorption spectrophotometry. *At. Abs. Newsletter* 6 (1967) 1.

67. SCHOLES, P. H., The application of atomic-absorption spectrophotometry to the analysis of iron and steel. *Analyst* 93 (1968) 197.

68. PRICE, W. J., and ROOS, J. T. H., The determination of silicon by atomic-absorption spectrophotometry, with particular reference to steel, cast iron, aluminium alloys and cement. *Analyst* 93 (1968) 709.

SILICATE ANALYSIS

69. BELCHER, C. B., and BROOKS, K. A., The determination of strontium in coal ash by atomic absorption spectrophotometry. *Anal. Chim. Acta* 29 (1963) 202.

70. BILLINGS, G. K., and ADAMS, J. A. S., The analysis of geological-materials by atomic absorption spectrometry. *At. Abs. Newsletter* 3 (1964) 65.

71. TRENT, D. J., and SLAVIN, W., Determination of the major metals in granitic and diabasic rocks by atomic absorption spectrophotometry. *At. Abs. Newsletter* 3 (1964) 17.

72. TRENT, D. J., and SLAVIN, W., Determination of various metals in silicate samples by atomic absorption spectrophotometry. *At. Abs. Newsletter* 3 (1964) 118.

73. BILLINGS, G. K., The analysis of geological materials by atomic absorption spectrometry: II Accuracy tests. *At. Abs. Newsletter* 4 (1965) 312.

74. BURRELL, D. C., An atomic absorption method for the determination of cobalt, iron and nickel in the asphaltic fraction of recent sediments. *At. Abs. Newsletter* 4 (1965) 328.

75. SLAVIN, W., The application of atomic absorption spectroscopy to geochemical prospecting and mining. *At. Abs. Newsletter* 4 (1965) 243.

76. Passmore, W., and Adams, P. B., Determination of iron and zinc in glass by atomic absorption spectrophotometry. *At. Abs. Newsletter* 4 (1965) 237.

77. Jones, A. H., Analysis of glass and ceramic frits by atomic absorption spectrophotometry. *Anal. Chem.* 37 (1965) 1761.

Cement Analysis

78. Sprague, S., Cement Analysis. *At. Abs. Newsletter* 2 (1963) 43.

79. Takeuchi, T., and Suzaki, M., The determination of sodium, potassium, magnesium, manganese and calcium in cement by atomic absorption spectrophotometry. *Talanta* 11 (1964) 1391.

80. Capacho-Delgado, L., and Manning, D. C., The determination by atomic absorption spectroscopy of several elements including silicon, aluminium and titanium in cement. *Analyst* 92 (1967) 553.

81. Price, W. J., and Roos, J. T. H., The determination of silicon by atomic absorption spectrophotometry, with particular reference to steel, cast iron, aluminium alloys and cement. *Analyst* 93 (1968) 709.

Soil Analysis and Fertilizers

82. David, D. J., The determination of exchangeable sodium, potassium, calcium and magnesium in soils by atomic absorption spectrophotometry. *Analyst* 85 (1960) 495.

83. Allan, J. E., The determination of zinc in agricultural materials by atomic absorption spectrophotometry. *Spectrochim. Acta* 17 (1961) 467.

84. David, D. J., The determination of strontium in biological materials and exhangeable strontium in soils by atomic absorption spectrophotometry. *Analyst* 87 (1962) 576.

85. McBride, C. H., Determination of minor nutrients in fertilizers by atomic absorption spectrophotometry. *At. Abs. Newsletter* 3 (1964) 144.

86. Nadirshaw, M., and Cornfield, A. H., Direct determination of manganese in soil extracts by atomic absorption spectroscopy. *Analyst* 93 (1968) 475.

Trace Metals in Water and Effluents

87. Butler, L. R. P., and Brink, D., The determination of magnesium calcium, potassium, sodium, copper and iron in water samples by atomic absorption spectroscopy. *S. African Ind. Chemist* 17 (1963) 152.

88. SPRAGUE, S., and SLAVIN, W., Determination of very small amounts of copper and lead in KCl by organic extraction and atomic absorption spectrophotometry. *At. Abs. Newsletter* 3 (1964) 37.

89. DELAUGHTER, B., The determination of sub-p.p.m. concentrations of chromium and molybdenum in brines. *At. Abs. Newsletter* 4 (1965) 273.

90. PLATTE, J. A., and MARCY, V. M., Atomic absorption spectrophotometry as a tool for the water chemist. *At. Abs. Newsletter* 4 (1965) 289.

91. BIECHLER, D. G., Determination of trace copper, lead, zinc, cadmium, nickel and iron in industrial waste waters by atomic absorption spectrometry after ion exchange concentration on Dowex A-1. *Anal. Chem.* 37 (1965) 1054.

92. MAGEE, R. J., and RAKMAN, A. K. M., Determination of copper in sea water by atomic absorption spectroscopy. *Talanta* 12 (1965) 409.

93. FISHMAN, M. J., The use of atomic absorption for analysis of natural waters. *At. Abs. Newsletter* 5 (1966) 102.

94. JAYNES, T., and FINLEY, J. S., The determination of manganese and iron in sea water by atomic absorption spectrometry. *At. Abs. Newsletter* 5 (1966) 4.

PETROLEUM ANALYSIS

95. ROBINSON, J. W., Determination of lead in gasoline by atomic absorption spectroscopy. *Anal. Chim. Acta* 24 (1961) 451.

96. SPRAGUE, S., and SLAVIN, W., Determination of the metal content of lubricating oils by atomic absorption spectrophotometry. *At. Abs. Newsletter* 2 (1963) 20.

97. TRENT, D. J., and SLAVIN, W., The direct determination of trace quantities of nickel in catalytic cracking feedstocks by atomic absorption spectrophotometry. *At. Abs. Newsletter* 3 (1964) 131.

98. BURROWS, J. A., HEERDT, J. C., and WILLIS, J. B., The determination of wear metals in used lubricating oils by atomic absorption spectroscopy. *Anal. Chem.* 37 (1965) 579.

99. MEANS, E. A., and RATCLIFF, D., Determination of wear metals in lubricating oils by atomic absorption spectroscopy. *At. Abs. Newsletter* 4 (1965) 174.

100. Sprague, S., and Slavin, W, A rapid method for the determination of trace metals in used aircraft lubricating oils. *At. Abs. Newsletter* 4 (1965) 367.

101. Trent, D. J., The determination of lead in gasoline by atomic absorption spectroscopy. *At. Abs. Newsletter* 4 (1965) 348.

102. Kerber, J. D., The direct determination of nickel in catalytic-cracking feedstocks by atomic absorption spectrophotometry. *Appl. Spectroscopy* 20 (1966) 212.

103. Moore, E. J., Milner, O. I., and Glass, J. R., Application of atomic absorption spectroscopy to trace analyses of petroleum. *Microchem. J.* 10 (1966) 148.

104. Wilson, H. W., Note on the determination of lead in gasoline by atomic absorption spectrometry. *Anal. Chim. Acta* 38 (1966) 921.

105. Mostyn, R. A., and Cunningham, A. F., Some applications of atomic absorption spectroscopy to the analysis of fuels and lubricants. *J. Inst. Petrol* 53 (1967) 101.

VI

Characteristics of Standard Equipment

ONE of the more time consuming, exacting, and sometimes frustrating duties that fall to the lot of a chief chemist is that of purchasing expensive analytical instruments. This responsibility is not always facilitated by the sales publications of the manufacturers. For this reason it is felt that a basic description of the main features of commercial equipment, and the points a chemist should consider in choosing a unit for his own uses, may be of interest.

There are several ways of choosing an atomic absorption spectrophotometer. One method, not infrequently employed, is that of selecting the most expensive unit within the laboratory budget and hoping that it will automatically possess a capability in excess of that required. This approach can often produce unlooked-for complications. The equipment, even if the senior analyst is willing to allow junior staff to use it, can be so complicated to operate that only specially trained personnel can do so. It can also be slower in operation than less expensive equipment!

Desirable Features of an Atomic Absorption Spectrophotometer

In attempting to assess the relative merits of various commercial units for use in a particular laboratory the chemist should consider what features he requires the instrument to possess. It is advisable firstly to consider the characteristics and concentration levels of the elements to be determined, together with the accuracy required. For low level work it is safe to say that all British, American and Australian production units present scale expansion facilities either built in or as modular accessories.

The dramatic improvements that have been made since the end of 1966 in the manufacture of hollow-cathode lamps have given instruments built around simple optics a performance comparable to their more elaborate counterparts.

Secondly it is important to take account of the ability and experience of the operators who are to use the equipment. Obviously if the unit is to be used by junior staff for routine work, or by personnel not specifically trained in the disciplines of analytical chemistry and instrument engineering, it should be simple and reliable in operation.

Speed of operation is always of importance, and generally the simpler the equipment the faster will it be to operate.

Versatility is a final important feature. Most commercial units can be operated as atomic absorption spectrophotometers and flame emission units. Some can be adapted also for atomic fluorescence. It is when a unit is to be used for high-temperature flame emission or atomic fluorescence that it is essential that it should incorporate a high resolution monochromator, possessing also high light-gathering capacity.

Basic Features of Standard Instruments

All atomic absorption spectrophotometers consist of:

1. a nebulizer system
2. a flame system
3. an optical system, comprising lamps, single or double beam light passage and monochromator
4. a photomultiplier
5. an electronic readout system.

It is in the components and assembly of these essential units that differences in performance and price (especially the latter) occur.

Nebulizers for Premix Burners

All of these operate on the same principle. The efficiencies of such nebulizers are rather low; only about 15 per cent (at maximum) of the solution aspirated is nebulized into the fine droplets (Plateau's spherules) that are carried to the flame of a premix burner. It is in this sector that improvements in the performance of atomic absorption equipment are most needed. At present the capabilities of all commercial nebulizers are very similar.

Flame System

There are two types: turbulent flow or total consumption burners and laminar flow or premix burners.

TURBULENT FLOW BURNERS

In construction this type of burner consists of three concentric tubes, the inner one being a fine capillary which carries the sample into the flame, by means of the suction created by the passage of gases in the surrounding tubes. The sample is injected into the flame at a rate which is dependent upon the flow of gas and also the viscosity of the sample.

Thus in general this type of burner is characterized by the fuel gas and supporting gas being unmixed until they reach the base of the flame. The solution to be nebulized is also introduced at this point and thus the burner and nebulization system are built as a single unit. These burners are often known as 'direct injection' or 'total consumption' burners, since all the sample liquid enters the flame and is converted to spray at the point of entry. At first sight such a system might appear to offer 100 per cent efficiency; and cause one to question why premix burners are used at all.

In practice it is found that total consumption burners must be constructed in a manner similar to an oxy-acetylene blow torch, i.e. they can only have a very short path length.

Secondly the droplets aspirated to the flame are of widely varying sizes, so that the processes of evaporation and thermal decomposition occur at virtually all heights in the flame; i.e. it is impossible to select a particular height at which an overwhelming majority of the absorbing species will be found.

The turbulent nature of the flame, in which the gases do not mix until they start to burn, gives rise to a high noise level. The shortness of the light path through the flame, and the non-uniformity of droplet size, reduce sensitivity, and further aggravate the poor signal-to-noise ratio.

Attempts are made in some instruments to overcome these disadvantages by arranging for the light beam to make multiple passes through the flame. Apart from complicating the optical system and adding to the expense of the unit, such a set-up is intrinsically inefficient. The carrier gas/fuel ratio is widely different at varying heights in the flame, so that the absorption, and in extreme cases interferences, vary at each of the different light paths. Also a sensible percentage of the light is dissipated at the reflecting surfaces so that wider monochromator slit widths and higher photomultiplier gain have to be used, thereby increasing the background noise and possibly reducing sensitivity.

Laminar Flow Burners or Premix Burners

In this type of combustion system the supporting gas nebulizes the sample solution into an expansion chamber, where the larger droplets (constituting 85–90 per cent total aerosol) settle out and go to waste. The remaining portion of the aspirated solution, which is in the form of a mist, of small droplets of reasonably uniform size, is carried into the burner barrel where it is pre-mixed with fuel gas (in some designs the fuel gas is added in the expansion chamber). The pre-mixed gases travel up the burner barrel and are combusted either on an array of holes or a slot. A slot is preferable for high-burning-velocity gas mixtures, e.g. air–acetylene or nitrous oxide–acetylene.

With air–acetylene, air–hydrogen and air–propane, most commercial burners have flame paths 10 or 12 cm long. However, the slits of nitrous oxide–acetylene burners, because of the higher burning velocity of this gas mixture, are usually 5 cm in length.

Premix burners of this construction, by virtue of the length of the flame path give good sensitivity, and because non-uniformities in the flame itself tend to average themselves out over its length, acceptable signal-to-noise ratios are achieved.

Finally, because the droplets are of more uniform size, it is possible to select a definite portion of the flame where an overwhelming percentage of the absorbing species will be found. This allows much more reproducible, sensitive and in some cases interference-free determinations to be performed than with a total-consumption system.

The correct procedure to ensure optimum and uniform flame conditions is to select a burner height, set the oxidizing gas (air or nitrous oxide) to a fixed flow rate, and then to regulate the fuel so that peak absorption is obtained when a suitable standard solution is aspirated into the flame. If this procedure is used it will be found that for most elements there is a very wide range of burner heights over which the maximum sensitivity attainable is sensibly identical. This point is mentioned because so much misleading over-emphasis has been placed on the importance of burner height adjustment in some literature.

The most valuable advantage obtained from burner height adjustment is that it allows measurement to be made low in the flame, where noise due to flame fluctuations is minimized. This can be of considerable importance for low-level estimations.

Optical System

Lamps

Hollow-cathode lamps are the main spectral sources. There has been a steady increase during the last four years in the number of elements for which reliable hollow-cathode lamps can be fabricated, and in the light-output intensity from such devices. The production of the so called 'High-Spectral-Output' lamps (not to be confused with high-intensity lamps, see later) incorporating shielded cathodes, improved electrode assemblies and better choices of filler gases, now permits even the simplest atomic absorption spectrophotometers to select more absorbing wavelengths and gives them a performance comparable to more elaborate and expensive units.

For example the initial type of argon filled iron lamp gave a very low output intensity. This necessitated the use of fairly wide monochromator slit settings so that radiation not only of the most absorbing line at 2483 Å but also that from the less absorbing line at 2488 Å together with the non absorbing background was focussed on to the photomultiplier. An insensitive and grossly non-linear response curve was thereby produced. An instrument that had superior light-gathering power and resolution would obviously perform analyses better, with such a low power spectral device, than a unit that had a less efficient monochromator.

The spectra from the currently available neon filled lamps is not only more powerful but the relative intensity of the most absorbing line at 2483 Å to the less absorbing lines around 2488 Å has been improved. The improved output from this newer type of source allows narrow monochromator slit settings to be used, so that an instrument having a lower resolution monochromator approaches the analytical capability of one designed around more powerful optics.

Chromium provides an example where replacement of argon by neon as filler gas gives rise to a 'cleaner' spectrum, so that the analytical capability of simple standard equipment is enhanced.

For nickel (see below) the output intensity of lamps was firstly improved by substituting neon for argon as filler gas, and the spectrum was rid of the unwanted ionic line at 2316 Å by restricting the filler gas to a low pressure.

Perhaps the most valuable advance in this series of improvements was the production of hollow-cathode lead lamps that permit reliable use of the very sensitive 2170 Å line with simple standard atomic absorption spectrophotometers.

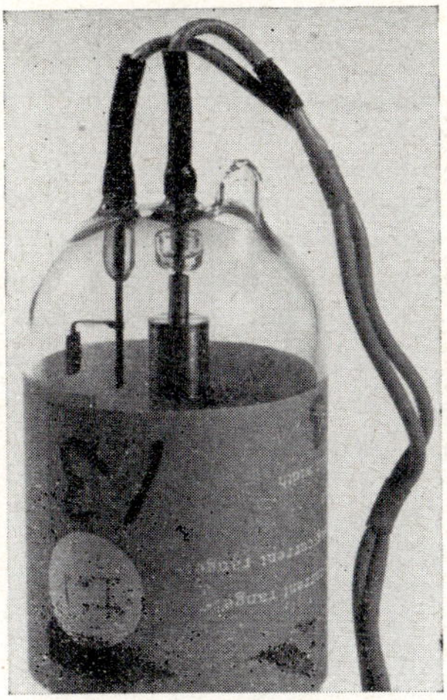

Fig. 6.1.
Older type, 'open structure' hollow-cathode lamp

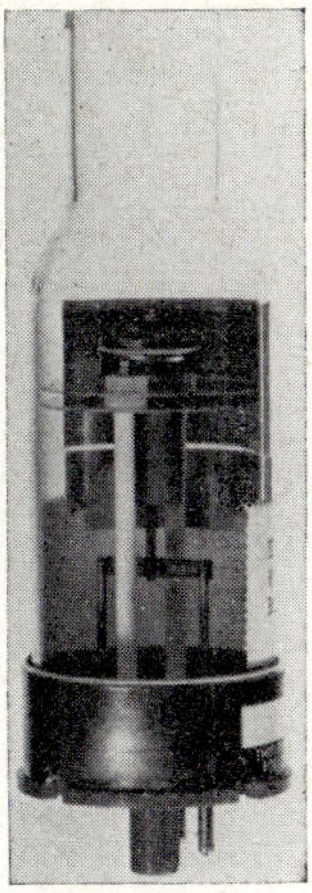

Fig. 6.2
'High spectral output' lamp

High Intensity Hollow-cathode Lamps (*also called 'High Brightness'*). These devices were developed by SULLIVAN and WALSH[1] and are essentially an open-structure hollow-cathode lamp containing two auxiliary electrodes capable of emitting a sensible population of electrons when electrically heated. The emitted electrons are designed to pass as a booster current across the hollow cathode. The energy available from the electrons of the booster current is low, so that the easily excited ground-state lines of the metal atoms sputtered from the cathode are preferentially enhanced, and the non-absorbing ion lines are reduced in intensity. Because of this enhancement of the ground-state atomic resonance lines, and suppression of unwanted ion and filler gas lines, it was anticipated that these high intensity devices would permit more linear, sensitive and stable atomic absorption measurements to be made. For some elements indeed, notably nickel, this aim was realised.

High intensity lamps, though, have a number of serious practical disadvantages, and of the two companies that originally produced them, one has virtually abandoned their manufacture. Briefly, these lamps require an extra power supply to provide the high auxiliary current; they are costly, their lifetimes are short, and they can only be produced for certain elements.

Finally, the drastic improvements in performance that have been made in standard hollow-cathode designs have all but rendered the high-intensity types superfluous.

Fig. 6.3 High intensity lamp, showing auxiliary electrodes

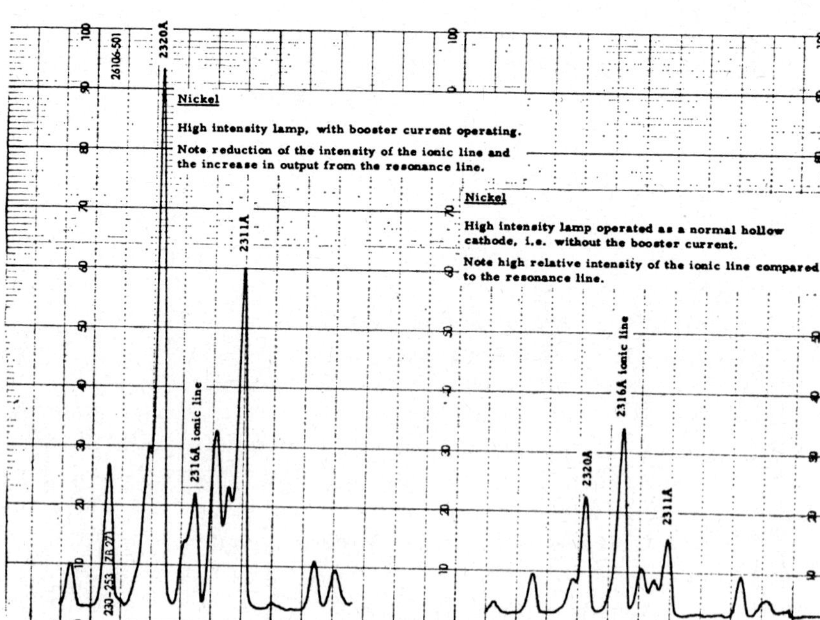

Fig. 6.4

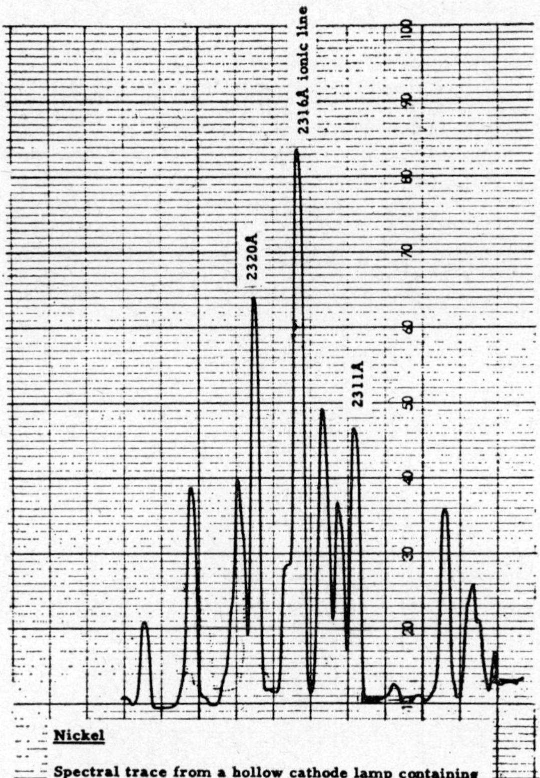

Nickel

Spectral trace from a hollow cathode lamp containing too high a pressure of filler gas.

Note the very high relative intensity of the ionic line.

Fig. 6.5

These points are more specifically illustrated by the data for nickel reproduced in Fig. 6.4 to 6.6.

Fig. 6.4 shows the spectrum obtained from a high intensity lamp operated with and without the booster current. Without this auxiliary current, i.e. with the source operated as a normal hollow-cathode lamp, not only is the output much less intense than when used in the true high intensity mode, but the non-absorbing ionic line at 2316 Å is more prominent than the resonance line at 2320 Å.

The ionic radiation is due to collision of filler gas atoms with nickel atoms, and this fact is illustrated in Fig. 6.5 and 6.6. Fig 6.5 shows the spectrum for a normal hollow-cathode lamp containing too high a pressure of filler gas. Fig. 6.6 shows a spectral trace for a second standard hollow-cathode lamp, that contains neon at a lower pressure than that in the lamp for which the previous trace was produced. It illustrates the

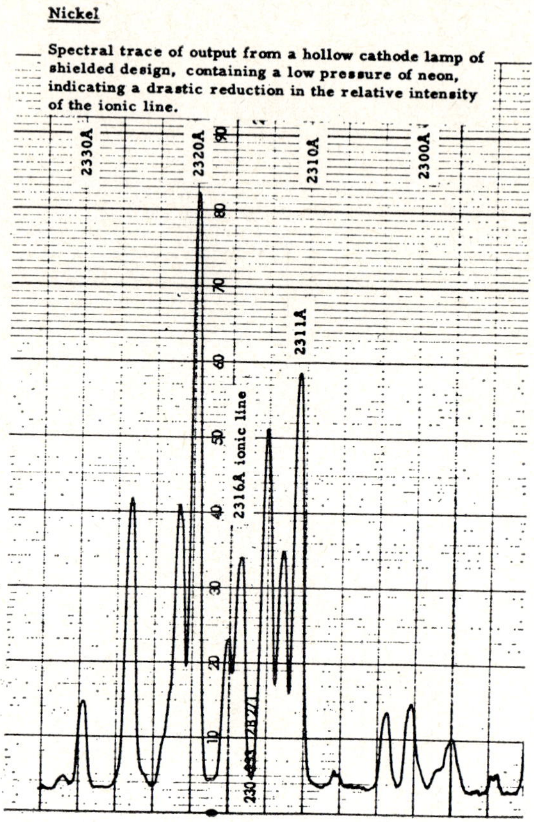

Fig. 6.6

fact that undesirable ionic radiation can be reduced to an insignificant level by controlling filler gas pressure, thereby enabling a normal hollow-cathode lamp to be constructed for nickel that possesses all the essential benefits of a high intensity lamp. With such a device an instrument designed around a low resolution monochromator provides an analytical performance similar to that of a unit having high resolution optics.

Microwave, or Electrodeless Discharge Tubes. (E.D.T.)[2,3]. The production and operation of these devices have been described in detail by West *et al.* Essentially they consist of a sealed quartz vial, about 0·8 cm internal diameter, 4 cm long and wall thickness 1·1 mm, containing a small quantity of the required element, usually as the iodide, under a pressure of between 1 and 4 mm of argon. When placed in a high-frequency electrostatic field, produced by a microwave generator, they emit energy of the characteristic wavelengths of the material they contain.

The general advantages and disadvantages of electrodeless discharge tubes with respect to hollow-cathode lamps are listed below.

Microwave-excited electrodeless discharge tubes retail for about half the price of hollow-cathode lamps and possess exceedingly long lifetimes. They can even be manufactured by a competent technician in a laboratory equipped with high vacuum and glass blowing facilities. They emit very sharp spectral lines of much higher intensity than can be obtained from a hollow-cathode lamp.

Electrodeless discharge tubes suffer from the disadvantage that they require for their operation a special microwave generator, which is a fairly costly piece of equipment. More seriously, at their present state of development the tubes vary rather widely in characteristics and are subject to more operating parameters than are hollow-cathode lamps. To be specific, the performance obtainable from an electrodeless discharge tube depends upon the position of the tube in the microwave cavity, the extent of cooling to which the cavity is subject and the operating power. They require a very long warm-up period to reach stability of output, at least 30 minutes, and in extreme cases over 2 hours. The variations in output that occur over this period are usually towards higher intensity in the early stages, but can later show a drop-off in energy. This phenomenon, it is thought, is due to the vapour

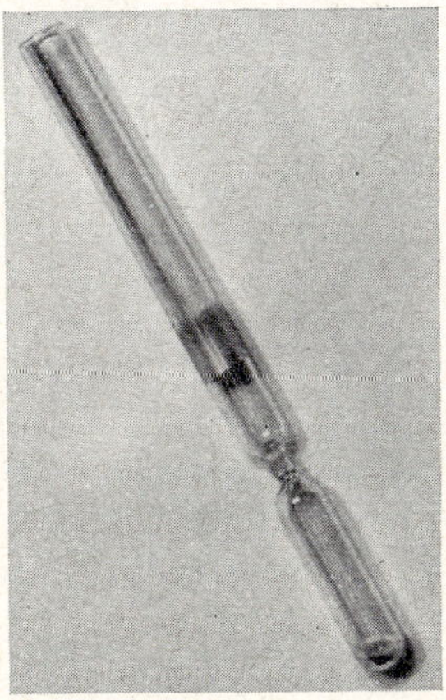

Fig. 6.7
Microwave discharge tube

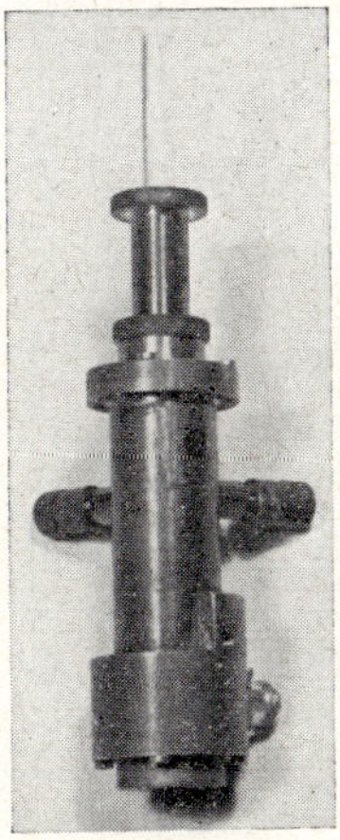

Fig. 6.8
Microwave discharge cavity with tube in position

which gives rise to the discharge physically moving around inside the tube.

The hollow-cathode lamps that are currently being produced require only about 7 to 15 minutes to reach stability of output. They also emit sharp spectral lines, and are subject to only one operating parameter, lamp current.

For these reasons it must be admitted, at the present time, that electrodeless discharge lamps cannot generally be recommended as a substitute for hollow-cathode lamps in the routine laboratory, where reliability, speed, and convenience of operation are important.

The Single and Double Beam Systems

Most commercial atomic absorption spectrophotometers use a single beam optical system. With this system, as its name implies, the light from the lamp is simply passed through the flame, as a single beam, to the monochromator, and focussed on to the photomultiplier. Units incorporating this principle are simple in construction and easy and quick to operate.

With the double-beam system the light from the source is split, usually by means of a rotating sector mirror into two beams, one of which passes through the flame, while the second (the reference beam) is deviated around it. At a point beyond the flame the two beams are recombined and their ratio is electronically compared.

The advantages claimed for such a system are:

(i) That it is possible to take readings directly a lamp is switched on, instead of waiting for it to reach stable maximum emission. As this warming period is only about 5 to 15 minutes and virtually all standard instruments are now offered with modular preheating units or turrets, this is not a very important 'advantage'.

(ii) The second advantage quoted is that the system corrects for small variations in intensity of light output during a run, so that a steady base line is obtained. This may have been of some significance in the early days of atomic absorption, but the lamps now produced are so stable that this advantage is not apparent for most determinations.

For very prolonged runs using a continuous sampling procedure, for example the determination of calcium in blood serum in say 100 or more samples that are selected, prepared, diluted and fed automatically to an atomic absorption spectrophotometer, the ability of a double-beam instrument to correct for zero drift could be advantageous. It so happens, however, that calcium lamps are more prone to variations in output over long running periods than lamps for most of the other elements.

The double beam system, though, possesses some shortcomings.

(i) The reference beam corrects only for the variations in lamp output, which are negligible. It does nothing to overcome the most serious source of instability, fluctuations in flame background.

(ii) The utilization of the light from the lamp is much less efficient with a double beam system than with a single beam, because a high percentage of the energy available is lost by sharing between the two beams, and on the more numerous optical surfaces required. This means that wider slit settings have to be employed so that unless a high resolution monochromator of good light gathering capacity is used, a decrease in sensitivity can result.

(iii) A manual-balance double-beam instrument is somewhat slower to operate than one using a single beam, because it is necessary to balance the reference beam before taking every reading.

THE MONOCHROMATOR

Monochromators depend on either a prism or a diffraction grating for their operation. The prisms vary in performance according to their composition and quality of manufacture. Grating performances vary mainly according to size and the closeness of their lines. The various systems used for assembling the components of a monochromator, for example LITTROW, CZERNY-TURNER etc., also affect performance.

The terms used in describing monochromators, the general features of these important components, and factors affecting their performance are discussed briefly below.

The two optical characteristics that influence the performance of an atomic absorption spectrophotometer are (a) the light intensity that the monochromator can focus on to the photomultiplier, and (b) the ability of the monochromator to separate lines that are close together in the spectrum. The factors that affect the intensity of light focussed on the photomultiplier are *f*-number and efficiency of light utilization

f-number. The brightness of an image formed by an optical instrument. or in the case under consideration, the light intensity that can be focussed upon the photomultiplier, is governed by the *f*-number, which is given by the ratio f/a, where f is the focal length of the instrument and a is the diameter of the effective entrance pupil. The lower the *f*-number the brighter is the image, so that for a given focal length, it is advantageous if the value of a is made as large as possible. This can be accomplished by increasing the physical size of the internal optics of the monochromator, i.e., a unit containing large mirrors and a large grating, or prism, will transmit more light than one of equal focal length, containing smaller components.

Efficiency of Light Utilization (Prism versus Grating). Light dispersed into its component wavelengths by a prism is all of one order; whereas with a diffraction grating it is shared between the several orders produced. From this point of view, therefore, a prism should be capable of transmitting a higher intensity of light to the detector. In practice the angles at which gratings are blazed can be adjusted so as to concentrate about 90 per cent of the light into a particular order.

The factors that govern the separation of spectral lines are dispersion, and resolution.

Dispersion. Angular dispersion is defined by the ratio $d\theta/d\lambda$, i.e. it is the rate of change of the angle of deviation with change of wavelength. A prism produces dispersion due to the fact that the refractive index of its material of composition is dependent upon wavelength. The refractive index always increases as the wavelength decreases, i.e. the violet end of the spectrum is deviated more than the red end. The most important characteristic of the dispersion produced, is that it increases with decreasing wavelength, so that the violet end of the spectrum, produced by a prism, is spread out on a larger scale than the red-end. This means that with a monochromator utilizing a prism as its basic component, the wavelength scale of the spectrum produced is compressed towards the red end.

The angular dispersion produced by a grating (in a particular order) is inversely proportional to the ruling separation (or grating space), i.e. the closer the lines are together, the more widely spread will be the spectra produced. More important, though, the various spectral lines produced differ in angle by amounts that are directly proportional to wavelength differences. This simple linear relationship for the lines produced by a grating is one of the chief advantages of a grating over a prism monochromator.

In practice, it is the *linear dispersion* of the spectral lines that control the separations possible with the monochromator. The value of this function (at the focal point) is calculated by multiplying the angular dispersion by the focal length of the instrument. Thus the greater the focal length the better will be the linear separation of the various spectral lines.

The reciprocal of the linear dispersion is called the reciprocal dispersion, and gives the spread of the spectrum in Å per mm at the focal point of the system. From this figure the band pass, i.e. width of the spectrum in Å that will be focussed upon the detector at various slit settings, can be computed.

Resolution. The resolution, or resolving power of an optical instrument defines its ability to produce separate images of objects that are very close together. It is given by the ratio $\lambda/d\lambda$, where $d\lambda$ is the smallest wavelength difference that produces resolved images. With a grating the diffraction pattern sets a theoretical upper limit to this characteristic, and it can be shown that (in a given order) it is proportional to the total number of rulings, but is independent of spacing.

For a prism it increases with the refractive index and the length of the base.

With a grating monochromator, a spectrum formed in a low order of diffraction is always used, so as to avoid complications due to overlapping orders. To obtain adequate dispersion, in a low diffraction order, the grating space must be made as small as possible. At the same time the resolution will improve as the number of rulings is increased.

A 2-inch grating ruled at 15000 lines per inch has a resolution given by

$$R = \frac{\lambda}{d\lambda} = mN$$

where m is the diffraction order and N is the *total Number* of rulings on the grating.

i.e.

$$R = 2 \times 15000 = 30000 \text{ (for the first order)}$$

Thus at 2400 Å the smallest wavelength interval that can be resolved by the grating will be

$$d\lambda = \frac{2400}{30000} = 0{\cdot}08 \text{ Å}$$

Thus, from the point of view of optics, a further major advantage of a grating monochromator is not merely the good uniform dispersion attainable over the whole spectrum, but also the high resolving power, i.e. narrowness of the spectral lines produced.

To summarize, an instrument using a grating monochromator will possess the characteristics of uniform dispersion over the whole spectrum. It will certainly have superior dispersion compared with a prism instrument at high wavelengths, and will be at least as good at low wavelengths. Its resolution is likely to be at least as good at all wavelengths. Light utilization is theoretically better with a prism than with a grating since there is no sharing of energy between orders. At lower wavelengths, though, absorption by the prism material can reduce the energy transmitted. On the whole, there is more to recommend a grating instrument than one utilizing a prism monochromator, and most manufacturers recognize this fact.

Having selected the basic type of monochromator to give the best performance, the questions of focal length, physical size of optics, and grating ruling can be considered. The greater the focal length the better will be the linear dispersion, but at the same time the light energy transmitted will be reduced. Transmission can be improved by increasing the physical size of the components; this, and the narrowness of rulings on the grating, will also improve resolution. Thus a balance, which also controls the price of the instrument, has to be struck between focal length and component sizes that give a manageable sized unit, of acceptable linear dispersion, light transmission and resolution.

For atomic absorption determinations of most of the common metals a monochromator of moderate quality is suitable. However, for determinations of metals for which the lamps possess poor output and complex spectra, a better monochromator is necessary. Such elements, because of the vast improvements in lamp technology already commented upon, are now few and diminishing in number. It is in the fields of high temperature flame emission, and atomic fluorescence, that the superior capabilities of a high quality monochromator will be most needed.

Photomultiplier

Some photomultipliers used in commercial atomic absorption spectrophotomoters, together with their nominal characteristics, are tabulated below:

E.M.I.	9662B	Quartz window	Range 1850 Å–6600 Å
E.M.I.	9663B	u.v. glass window	Range 2000 Å–8000 Å
E.M.I.	9592B	u.v. glass window	Range 1900 Å–8000 Å
E.M.I.	6256B	u.v. glass window	Range 1850 Å–6500 Å
H.T.V.	R136	Quartz window	Range 1850 Å–8500 Å
H.T.V.	R106	Quartz window	Range 1850 Å–6600 Å
H.T.V.	R376	Quartz window	Range 1850 Å–8500 Å
R.C.A.	I.P.28	u.v. glass window	Range 1900 Å–8000 Å

There can be considerable variation in performance between tubes of the same nominal characteristics. For example, the H.T.V. R136 tubes, which are widely used in atomic absorption spectrophotometers, are all specially selected, firstly by the component manufacturer and then again by the instrument manufacturer.

Electronic Readout System

Modulation

In general, not only will there be light of a particular wavelength originating from the lamp, falling upon the photodetector, but also light of the same wavelength arising from the flame. It is necessary to distinguish between these two sources of radiation, since it is the measurement of that from the lamp only which is required. This requirement is attained by modulating the light from the lamp, with either a mechanical chopper, or by using an a.c. power supply and tuning the electronics of the detector to this particular frequency. The d.c. signals from the flame and other extraneous sources are thus eliminated.

The main advantage of electronic modulation over a mechanical chopper is that no moving parts are required. However, if a double-beam system using 'time sharing', i.e. one in which the light from the measuring and reference beams is passed onto the same photomultiplier, is used, then a mechanical chopper is essential.

The random background noise that originates from the flame and the transistors is always more noticeable at low frequencies. High-frequency photomultiplier noise increases above about 1000 Hz. The optimum modulation frequency is thus round about 350 to 400 Hz.

Elimination of Background Noise by Integration of the Signal

The two features that control limits of detection, and hence the accuracy, with which a determination can be performed are sensitivity and background noise. Diminution of background noise will improve the accuracy of all estimations, and will noticeably enhance the performance of an instrument for low level determinations. Fluctuations in readings can very simply be observed and averaged by feeding the signal to a chart recorder and drawing a line midway between the maximum height of the peaks and minimum of the valleys.

A neat and valuable capability of some instruments (notably the E.E.L. 240) is that provided by the inclusion of an electronic integrating device. This functions by allowing the total fluctuating current, over about 15 seconds, to charge a capacitor. The voltage attained by this unit is then applied to the read-out device to give a very steady signal. The integrating read-out of the E.E.L. 240 is capable also of scale expansion up to ×10. The signal from such a device is absolutely steady. Successive readings from a determination having a low background noise (e.g. Mg, Ni, Fe) are highly reproducible, those from intrinsically noisy estimations (Sn, Sb, etc.) are less reproducible, but as the individual readings are steady a reliable average can be computed.

Zero Supression

The determination of calcium in cement necessitates very accurate estimation of small differences of Ca between the 4 and 6 μg/ml levels. This accuracy can only be attained by scale expansion between these limits. It is essential for this determination (and others like it) that the instrument should be capable of expansion of selected portions of a scale, rather than of the whole range from zero upwards. This capability is known as zero suppression. Not all commercial instruments possess this ability, which is highly desirable if the instrument will be required to perform accurate determinations of macro constituents.

Chap. VI References

1. Sullivan, J. V., and Walsh, A., High intensity hollow cathode lamps. *Spectrochim. Acta* 21 (1965) 721.
2. Dagnall, R. M., Thompson, K. C., and West, T. S., Microwave-excited, electrodeless discharge tubes as spectral sources for atomic fluorescent and atomic absorption spectroscopy. *Talanta* 14 (1967) 551.
3. Dagnall, R. M., and West, T. S., Some applications of microwave-excited, electrodeless discharge tubes in atomic spectroscopy. *Applied Optics* 7 (1968) 1287.

VII

Some Further Techniques

THE information in the preceding chapters, it is hoped, will enable an analyst to develop and perform reliable estimations by atomic absorption spectroscopy. The subject, though, is one that tends to interest scientists to an extent that they enquire beyond the bounds of knowledge essentially required for their purposes. It is at this point that the interested newcomer to the field can start to feel confused about the hardware and procedures referred to at lectures and in original articles. This chapter is therefore devoted to a description of these developments, of which some are of mere historical interest, some may be incorporated into standard instruments in the future, and others have already found specialized practical application.

Nebulization

The ideal nebulizer would convert all the sample into a fine mist of uniform droplet size at a constant rate, so that a steady absorption measurement would be obtained. In doing so it would use exactly the amount of nebulizing gas required by the burner arrangement. It should be of simple construction, not easily blocked by particulate matter, made of corrosion-resistant materials, and would remain in adjustment for long periods. The performance should not be dependent on viscosity, surface tension, density, or volatility of the liquid.

In principle, a liquid stream emerging from an orifice is broken up into droplets by the atomizing gas flow. The liquid flow is normally produced by the action of the nebulizing gas. This means a given gas flow will produce a given pressure differential, so that the liquid flow, and hence the nebulizer performance, will depend on the physical characteristics of the liquid.

The break up into drops of a liquid emerging from an orifice has received much study. It is well known that a neck appears in the liquid jet, which finally breaks, to form a drop just in front of the neck, with the formation of an extra tiny drop (Plateau's spherule) at the neck itself. With a laminar flow burner the larger drops settle out in the expansion chamber, and only the fine mist enters the burner.

Ultrasonic Nebulization

When a beam of ultrasonic waves is passed through a liquid and directed at a gas interface, atomization of the liquid occurs. Liquid particles are ejected from the surface into the surrounding gas. Under proper conditions a very fine dense fog is produced.

The technique of ultrasonic nebulization has at least one advantage over conventional nebulization techniques; the fog particle size and fog density can be independently controlled.

With the pneumatic nebulizers normally employed in atomic absorption systems, reduction in particle size can only be effected at the expense of fog density, because to attain it the gas flow must be increased. In ultrasonic nebulization the fog density can be varied simply by adjusting the gas flow past the liquid surface. The amount of liquid suspended in gas is limited only by the rate at which it condenses out. The size of fog particles can be controlled by varying the frequency of the ultrasound. The higher the frequency the smaller the droplet. Lang[1] experimentally found the size of droplets to be

$$D = 0{\cdot}34\,\lambda$$

where

D = median drop diameter
λ = capillary wavelength produced by ultrasound.

$$\lambda = \left(\frac{8\pi\sigma}{\rho f^2}\right)^{1/3}$$

where

σ = surface tension
ρ = liquid density
f = ultrasonic frequency

The minimum gas flow required to transport the droplets must be in excess of their terminal velocity, which can be calculated from Stokes' Law.

Ultrasonic nebulizers were first used in conjunction with R.F.[2,3,4] plasmas, where low gas flow rates make the use of pneumatic nebulizers impracticable. Recently attention has been given to their use with conventional flames in atomic absorption spectroscopy[5,6]. The high efficiency of liquid sample to mist conversion, which is independent of gas flow rate, makes the system ideal in atomic spectroscopy. Also the range of droplet sizes produced is much narrower than with the normal mechanical nebulizer.

In the early design of ultrasonic nebulizers[3] sample changing was inconvenient. The sample was placed inside a tank, on to the base of which an ultrasonic transducer was bonded. The ultrasonic waves, produced by applying an 800 kHz signal, were focussed by a plano-concave leucite lens on to the liquid surface of the sample. As the R.F. output was increased to about 3 watts/cm^2 at the transducer, a stable fog was produced above the liquid surface, which could be transported over appreciable distances with little condensation of droplets. The average droplet size was about 5 microns

With a later design the sample was drip-fed on to the transducer at a fixed rate and total nebulization was achieved[7]. A much quicker sample changeover and wash through could be obtained with this system.

Types of Flame

Low Temperature Flames

Air–coal gas and air–propane flames have been used for the determination by A A S of certain elements, where the samples are in virtually pure aqueous media and chemical interferences are not present. The introduction of the air–acetylene flame in about 1958 to A A S improved determinations generally, and gave useful sensitivity for about 35 elements[8]. Even with this flame, though, several elements are still incompletely atomized, and chemical interferences are still severe for calcium, in the presence of phosphorus or aluminium.

Atomization, however, is not only a function of temperature, as shown by the absorptions of molybdenum and tin which exhibit higher values in the cooler fuel-rich air–acetylene flame than the hotter fuel-lean stoichiometric one. The use of air–hydrogen, although cooler, gives better results for selenium, tin and arsenic[9], but at its lower temperature interferences are more prevalent.

Premixed oxygen–hydrogen has such a high burning velocity that it has rarely been used for A A S. Also the turbulent oxygen–hydrogen flame used for flame emission spectroscopy is considerably reduced in temperature when aqueous samples are nebulized into it, and may only have a temperature of 2300°C[10] under these conditions. For this reason results obtained using this flame are less sensitive than might be expected, and chemical interferences still occur.

Table VII-1. BURNING SPEEDS AND FLAME TEMPERATURES

	Max. flame speed (cm. s^{-1})	Max. temp. (°C)
Air–coal gas	55	1840
Air–propane	82	1925
Air–hydrogen	320	2050
Air–50% oxygen–acetylene	160	2300
Oxygen–nitrogen–acetylene	640	2815
Oxygen–acetylene	1130	3060
Oxygen–cyanogen	140	4640
Nitrous oxide–acetylene	180	2955
Nitric oxide–acetylene	90	3095
Nitrogen dioxide–hydrogen	150	2660
Nitrous oxide–hydrogen	390	2650

HIGH TEMPERATURE FLAMES

Table VII-1 shows the burning veloc'ties and calculated flame temperatures for a number of gas mixtures which have been used or considered for A A S.

THE OXYGEN–CYANOGEN FLAME

The first published attempt to use this flame was made by ROBINSON[11,12], using a total-consumption burner in which the sample was aspirated using pre-mixed oxy–cyanogen. For several metals he achieved sensitivities slightly better than those obtained with a turbulent flow oxygen–hydrogen flame, but he was unable to detect any absorption for tin, tantalum, tungsten, and aluminium. Vanadium was only detectable with poor sensitivity. Since emission was obtained for these metals at the same wavelengths, in the above flame, lack of absorption signal may be explained by the emission being due to chemi-luminescence rather than thermal excitation.

Although the high temperature and low burning velocity of the oxygen–cyanogen flame looks promising for A A S, it suffers from the drawbacks that cyanogen is highly toxic and explosive, so it is unlikely to be accepted for routine use. Again, because of the poor diffusion of cyanogen into oxygen it is difficult to maintain a steady flame, and even small quantities of liquids aspirated can lower the temperature by as much as 2000°C[13].

Oxygen – Acetylene Flame

Because of its high burning velocity, work on atomic absorption was first reported using a turbulent burner[14], with a multi-pass arrangement, and a xenon arc as source. Slavin[15] and Manning, employing hollow cathode lamps, demonstrated strong absorption for aluminium, vanadium, titanium and beryllium.

The first published absorption for the lanthanides was reported in the oxygen–acetylene flame by Skogerboe and Woodriff[16]. Atoms produced in this flame were highly localized, and the replacement of water for the organic solvent originally used reduced the absorption to almost zero. Because of the inherent risk in using this flame little routine use has been found for it.

Nitrogen–Oxygen–Acetylene Flame

Amos and Thomas[17] investigated the atomic absorption of aluminium in a pre-mixed air–acetylene flame, in which the air was enriched with 50 per cent oxygen, burning on a stainless steel block 1·25 in thick, with a 3 cm × 0·45 mm slot. This flame, however, has some disadvantages for routine work, since it requires the provision of either a commercial O_2/N_2 mixture or a fairly complex gas mixing unit designed to prevent inadvertent use of oxygen-rich mixtures. Further, the high burning velocity of the gas mixtures required to produce atomization of metals forming refractory oxides limits the burner slit length to 3 cm.

Nitrous Oxide–Hydrogen Flame

There is little advantage gained in using this flame, although for the alkaline earth elements, which are hardly ionized in it, useful sensitivity and limits of detection have been found[18]. Molybdenum, germanium, beryllium and aluminium show negligible absorption in the nitrous oxide–hydrogen flame, although aluminium gives some atomic emission at 3942 and 3961 Å[19].

Nitric Oxide/Nitrogen Dioxide–Acetylene Flames

The lower burning velocity and higher temperatures attainable by burning acetylene with nitric oxide or nitrogen dioxide suggest the flames may be useful, but the corrosive nature of these oxides of nitrogen prevents their routine use.

SLAVIN[20] *et al.* have found that the sensitivity for several metals is slightly greater in the nitric oxide–acetylene flame than in the nitrous oxide–acetylene system, but the stability of the former flame is so poor that the detection limits are usually worse.

NITROUS OXIDE–ACETYLENE FLAME

This is a commonly used flame for atomic absorption work. Here the burning velocity is relatively low, and the high temperature of the flame is due in part to the energy liberated by the decomposition of the nitrous oxide. This flame can be burned on a 10 cm slot, although optimum sensitivity for most metals is obtained using a 5 cm slot.

AMOS and WILLIS[9] demonstrated that the flame could be used with commercial equipment, and found good sensitivity for almost all the metals which form refractory oxides and had proved difficult or impossible to atomize with the air–acetylene flame.

SEPARATED FLAMES

Pre-mixed hydrocarbon–air flames consist of two separate reaction zones: the primary reaction zone, where the combustible gas mixture burns principally to carbon monoxide, hydrogen and water, and the outer mantle, or secondary diffusion flame, where the hot gases burn with the atmospheric oxygen to carbon dioxide and water. The hottest part of these pre-mixed flames is normally in the centre of the flame just above the primary reaction zone.

In 1891 TECLU, and SMITHELS and INGLE[21] independently demonstrated the existence of the two zones, by mechanically separating the flame with a glass tube. The lower end of the tube was given an airtight fitting around the burner stem, and this prevented atmospheric oxygen reacting with the combustion products from the primary zone until they reached the top of the glass tube, where the secondary diffusion zone burned as before. In this way the space between the reaction zones, the interconal zone, is extended in length, and can be viewed without interference from the radiation of the secondary zone, which normally surrounds it in all unseparated flames.

It is the interconal zone which is normally used for analytical flame spectroscopy, and the potentiality of separated flames for flame emission and atomic absorption spectroscopy has been investigated by WEST, KIRKBRIGHT *et al.*[22–26].

SMITHELS and INLGE[21] reported the separation of the low-burning-velocity air-supported flames, coal-gas, ethylene, methane, etc., on an open Bunsen type burner port. KIRKBRIGHT, SEMB and WEST[22] reported the mechanical separation of an air–acetylene flame burning on a Méker-type head. A silica mechanical separator was used as this gave transmission of emission lines down to 2000 Å.

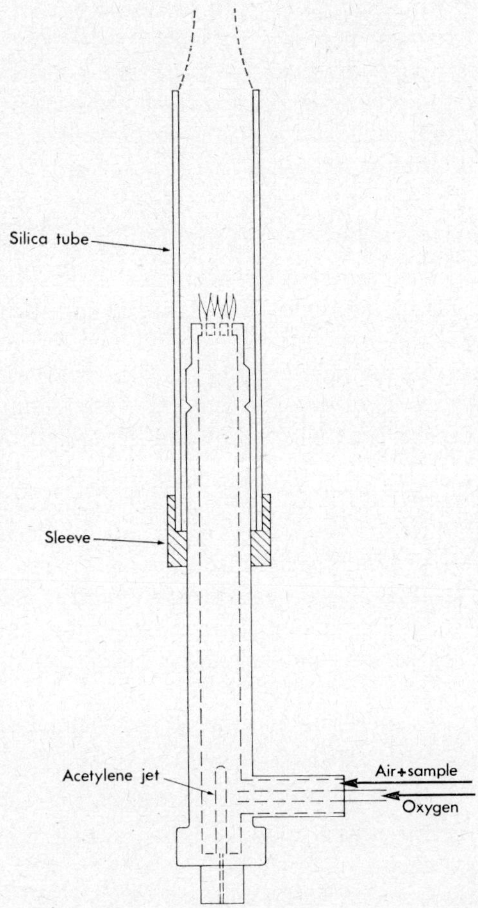

Fig. 7.1 Flame separation with a silica tube mechanical separator

The air–acetylene flame was separated (see Fig. 7.1) using a 12 cm long, silica tube of 20 mm internal bore, which was placed round the standard Unicam Méker-type burner, with an airtight sleeve enabling various separating distances to be obtained. Between 6 to 8 cm was found to be the most convenient distance from the top of the separator to the stainless steel burner head. The Unicam Méker head has 13 holes of about 1·2 mm diameter arranged in a 9 mm square pattern.

With this arrangement the flame is easily separated into a primary zone, consisting of 13 separate cones on the burner head, inside the silica tube, and a secondary diffusion zone burning at the top of the silica tube. Stable separated flames are obtained over a wide range of fuel/air ratios. The flame becomes turbulent with very lean mixtures and the use of very rich (luminous) mixtures leads to carbon forming on the inside walls of the separator. Turbulent noise occurs if a wide tube is used for separation, or if a separating distance greater than 10 cm is used with the 20 mm separator.

Emission Characteristics of Separated Air–Acetylene Flames

The primary zone of premixed air–acetylene flames radiates strongly over much of the visible and near-u.v. region of the spectrum, notably the molecular band emissions of CH, C_2, OH, and the strong CO continuum. This intense radiation, covering the larger part of the spectrum, makes the primary cone of little general use in flame spectroscopy. This emission is due to the CO continuum, generated by the reaction $2CO + O_2 \rightarrow 2CO_2 + h\nu$, and also the strong OH band system in the near-u.v. with peaks at 2811 Å and 3064 Å. The flame background due to CO and OH, that is observed when an unseparated flame is viewed, can be reduced by 2 orders of magnitude in the separated flame.

The noise level is also reduced considerably, and in emission studies considerable advantage is gained—particularly for elements whose emission lines lie in regions of very high background in the normal unseparated flame, e.g. bismuth (3068 Å).

A further effect of separation is to cool the flame because the heat normally provided by the reaction

$$CO + O \rightarrow CO_2 + 67{\cdot}6 \text{ kcal}$$

that occurs in the surrounding secondary diffusion zone has been removed from the interconal region.

In a later communication[23] Kirkbright, Semb, and West describe the mechanically separated nitrous oxide–acetylene flame for emission work. This also possessed a reduced CO and OH background. The nitrous oxide–acetylene flame possesses in the interconal region a red zone, due to the CN emission, which contributes to the highly reducing nature of the flame. When this flame is separated the CN emission is protected from the atmospheric oxygen and is thus extended. The burner used was in the from of a circular annulus 0·5 mm in width and 11 mm in diameter. This produced an intense primary reaction zone and red CN zone inside the separator, with the displaced secondary zone burning as a diffusion flame at the top of the silica tube.

The separated nitrous oxide–acetylene flame gave improved determinations of the refractory elements aluminium, molybdenum, and beryllium(27) by emission spectroscopy. The use of a long-path mechanically separated flame has been reported(24) for use in atomic absorption spectroscopy. This involved the use of the normal separated air–acetylene flame in conjunction with an electrically heated furnace. The interconal region was extended into a tube so as to obtain a long absorption path. Highly sensitive atomic absorption estimations of zinc, iron, copper, mercury, magnesium, and arsenic were reported.

The mechanical separator, however, eventually became corroded, and the transmission of the silica became progressively worse. The use of replaceable silica windows cemented to ground glass sockets gave the separator an increased life, but the advent of gas sheathing made it redundant.

The gas-sheathed flame was reported by HOBBS, KIRKBRIGHT, SARGENT, and WEST(26), the mechanical silica separator being replaced by a protective wall of nitrogen, flowing in a laminar annulus around the flame (see Fig. 7.2). This has the effect of 'lifting-off' the secondary zone, by preventing the access of atmospheric oxygen to support a diffusion zone in the lower parts of the flame.

This alternative method of separating the flame possesses the same advantages as the silica separator described earlier. The interconal zone exhibits low radiative background, so that greatly improved analytical signal to flame background ratios are obtained from aspirated metal ion solutions.

Several advantages are gained by gas sheathing over the mechanical silica separator. A wider range of fuel/air ratios may be used with safety and the analytical emission need not be viewed through the silica tube, which at low wavelengths becomes absorbing. The problem of etching or sooting of the inner surface of the separator is also eliminated. Again, the flame temperature above the primary cones is slightly reduced by separating, and from the electronic excitation temperature for iron using the two-line method, was found to fall from $2420 \pm 20°K$ to $2320° \pm 20°K$ for the air-acetylene flame.

The laminar nitrogen sheath was produced by passing nitrogen gas through a matrix of steel strip wound round the emission burner head in a spiral of alternate corrugated and flat 2 cm wide strip (0·1 mm in thickness).

Although silica separators could not easily be applied to long-path flames the gas sheathed arrangement can be. Five cm path length

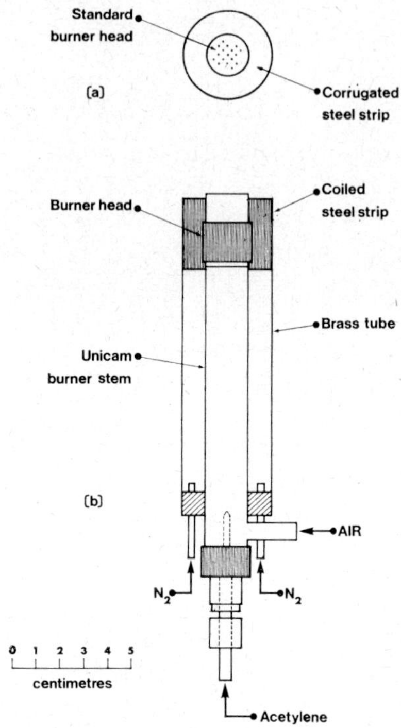

Fig. 7.2 Burner arrangement for nitrogen-separated air–acetylene flame

nitrous–acetylene and air–acetylene flames have been gas sheathed in a similar manner, and improved detection limits obtained for atomic absorption determinations.

The red CN zone in the nitrous oxide–acetylene flame is increased, due to less attack by atmospheric oxygen, and detection limits for refractory elements can generally be improved by a factor of 2 in the shielded flame.

Modified Burner and Nebulizer Systems

The Fuwa–Vallee Long Tube Burner[28,29,30]

This is a device for increasing the optical path of a total consumption burner system. The burner itself is inclined almost horizontally, and the flame directed into the end of a long ceramic tube. This tube is designed to confine the flame to the optical axis of the spectrophotometer, and thereby to combine the benefits of one hundred per cent

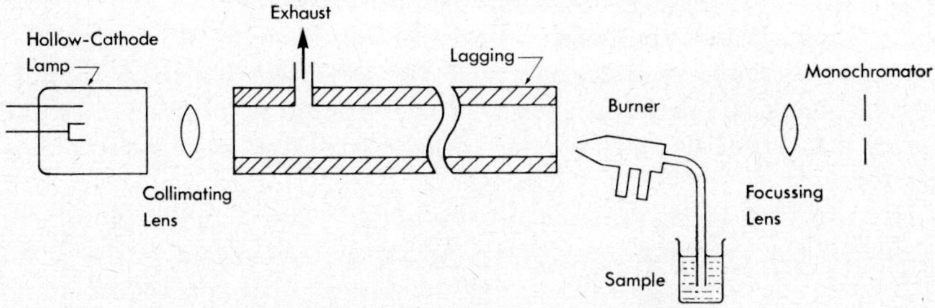

Fig. 7.3 Long tube burner arrangement

sample transfer to the flame and a long light-absorbing path. Improved detection limits, over those of other turbulent burners, are reported for metals (e.g. Pb, Zn, Cd) having easily dissociated oxides, but others such as Mg, Ca, Cr, Mo are converted to the stable oxide in the upper portions of the flame and, therefore, exhibit no improvement in sensitivity. The use of this device is rather untidy and of somewhat doubtful advantage, and is unlikely to be incorporated routinely into standard production instruments.

The Boling Three-slot Burner[31,32]

With the more elaborate atomic absorption instruments, which use fairly heavy gas flows, several advantages are claimed for this burner, which is designed to provide a flame that will more completely enclose the optical beam than a normal single-slot burner. It is stated that because the outer sheath of the flame is outside the light path, the background noise is diminished, while the sensitivity is increased since more light passes through the atomic vapour. It is further asserted that the fuel/air ratio is less critical.

Force-Feed Burner[33]

This is a modification of the total consumption turbulent flow burner. The sample, instead of depending on the suction produced by the flame gases, is mechanically fed into the base of the flame. This means that the viscosity of the sample does not control the aspiration rate, and different solvents may be fed into the flame at constant rates of flow. The liquid sample is broken up by the shearing action of the high-pressure gases issuing from the annular space round the capillary.

The Heated Mixing Chamber[34]

The object of this device is to pre-heat the fuel and carrier gases, so that the nebulized sample is flash dried into a fine aerosol of solid particles. This 'primary evaporation' process is thus completed before the sample reaches the flame, so that a more efficient dissociation into an atomic vapour should occur. It is also asserted that more particles are carried to the flame, at a higher concentration within the entraining gases. All these effects should combine to give rise to improved sensitivities.

In practice the equipment involved is rather cumbersome, and although the technique has been applied in one instrument it is not attractive to the majority of manufacturers. The procedure could possibly be dangerous if organic solvents are employed.

Methods other than with Flame

The procedures described below are of more limited application, and less convenient than the use of a flame, and cannot at present be regarded as a satisfactory basis for a standard analytical procedure.

The L'vov Furnace[35,40]

This device (see Fig. 7.4) is a micro furnace, the crucible of which is made in cylindrical form from graphite lined with tantalum foil. The ends are sealed with optical windows. A small solid sample (of about 100 μg) is introduced into the hollowed tip of an electrode that fits into the furnace midway along its length. The furnace is purged with argon. The sample is thermally dissociated by passage of an electric current, and the spectral measurements carried out by means of light, from a suitable source, focused down the length of the furnace tube.

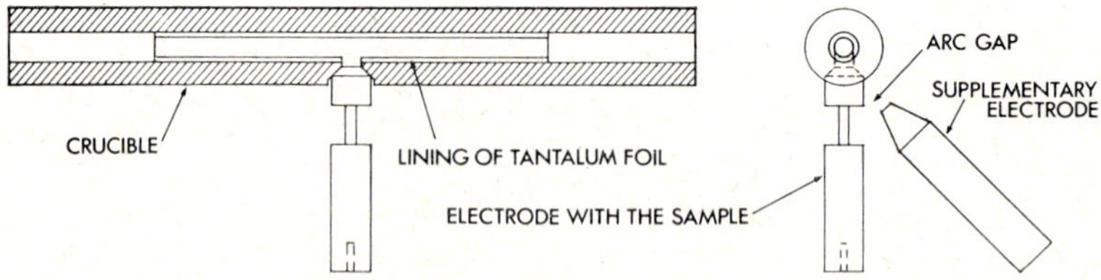

Fig. 7.4 The L'vov furnace

The perfection of this technique might, it is felt by responsible authorities, extend useful measurements into the far ultra-violet, improve the absolute sensitivity and reduce chemical interferences. Its overwhelming disadvantage so far is its very low precision.

THE SPUTTERING CHAMBER[36,37]

Cathodic sputtering (see Fig. 7.5) using a demountable hollow cathode, into which solid metallic samples could be inserted, was examined by GATEHOUSE and WALSH. The cathode was inserted into a cylindrical tube, containing a low pressure inert gas atmosphere. The atomic vapour produced was viewed through silica windows, fitted into the ends of the cylindrical chamber.
The procedure suffered from poor reproducibility and preferential sputtering of more volatile metals; nevertheless it was successfully exploited by GOLEB for the determination of uranium isotopes.

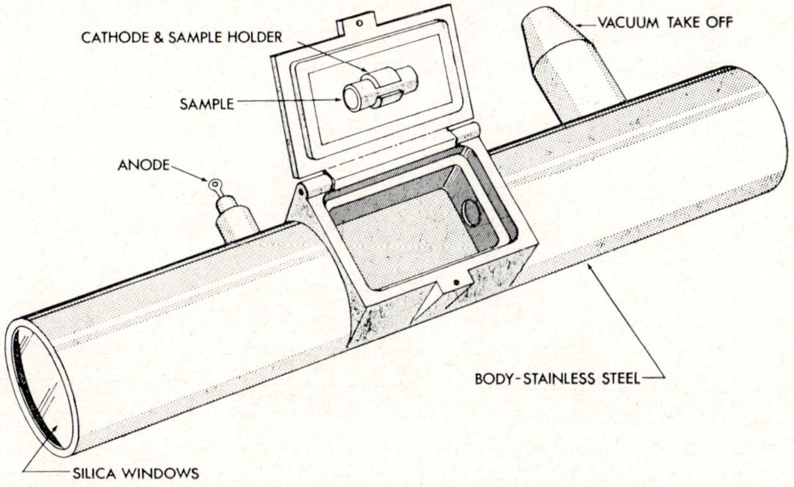

Fig. 7.5 A sputtering chamber

SOLID PROPELLANT—ATOMIZATION[38]

This method can be employed with some standard atomic absorption units. The sample is pulverized, mixed with the propellant powder and formed into a cylindrical pellet, which is slipped into a metal sleeve placed beneath the light beam of the atomic absorption spectrophotometer. The pellet is then ignited at its top surface, and is designed to burn uniformly, to give rise to a tall narrow column of flame that passes through the radiation beam, and lasts for sufficient time to obtain a stable absorption signal.

Standards are prepared either from previously analysed materials of similar matrix, or by wetting the propellant with a solution containing the required species and desiccating the mixture prior to forming the pellet.

The temperature attained by the pyrotechnic flame is about the same as that of the air–acetylene system, so the refractory metals cannot be estimated by the procedure. Applications are mainly restricted to the mining industry and, as might be expected, the analytical precision is of rather a low order.

Carbon Filament Atom Reservoir

This non-flame device described by West and Williams[39] uses the electrical resistance heating of a carbon filament to vaporize the sample. The sample in liquid form (about 5 μl) is placed on the carbon filament, supported between two electrodes in a small pyrex vessel, with transparent silica windows (Fig. 7.6). The chamber is purged with a small continuous flow of argon and the sample is vaporized by passage of about 100 A at 5 V through the filament, from a 10 kVA transformer unit. The filament reaches a temperature of about 2500°C within 5 seconds and the sample is vaporized. The whole unit is smaller than the average burner nebulizer used in A A S and, by placing a different glass cover over the electrode assembly, it can be instantly adapted for atomic fluorescence spectroscopy.

Although the unit may be operated under reduced, standard, or increased pressure conditions, the most reproducible results were obtained for a flowing system at atmospheric pressure. Argon was chosen, in preference to nitrogen as a purge gas to prevent oxidation of the filament, because it had no tendency to form cyanogen emission. In addition, less quenching of fluorescent emission occurs with monatomic argon than with diatomic nitrogen. A flow rate of 3·8 l/min was chosen to give a reasonable balance between sensitivity and precision.

Direct comparison with flame atomization was reported for silver and magnesium using A A S. Using the air–acetylene flame, a calibration for Ag of 1–10 ppm was obtained. If one assumes that a sample of 1 ml is sufficient for a determination this would be equivalent to 10^{-6}–10^{-5} g of silver. With the atom reservoir in place of the flame the range obtained was 10^{-9}–10^{-8} g, an improvement of three orders of magnitude. Similar improvements were obtained for magnesium.

This method of atomization is simple, and safe in operation. There is no danger of explosion and no great dissipation of heat. The filaments are self purifying in use and the cells show no measurable memory effect.

The complete absence of background absorption in atomic absorption spectroscopy work, or background emission in atomic fluorescence, make the carbon filament unit very attractive.

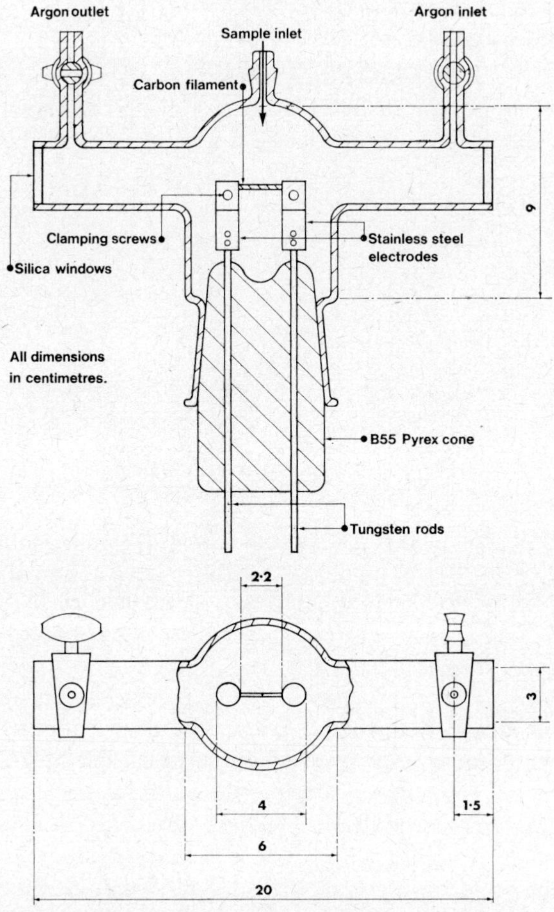

Fig. 7.6 Carbon filament atom reservoir

Modified Atomic Absorption Systems

The Resonance Detector [41,42]

In a conventional atomic absorption spectrophotometer the required spectral line is isolated by means of a monochromator. Walsh and his team at the C.S.I.R.O. in Melbourne have developed a technique in which the atomic resonance line(s) is isolated by exploiting the phenomenon of resonance. The principle is illustrated in Fig. 7.7.

Modulated radiation from a standard hollow cathode or high intensity lamp passes through the flame to a 'resonance lamp', in which the atomic vapour of the same element is produced by cathodic sputtering in a conventional unmodulated hollow cathode discharge lamp. The atomic vapour absorbs the resonance lines from the source, and some of this absorbed energy is re-emitted in all directions as resonance radiation. A portion, emitted at right-angles to the direction of the light beam

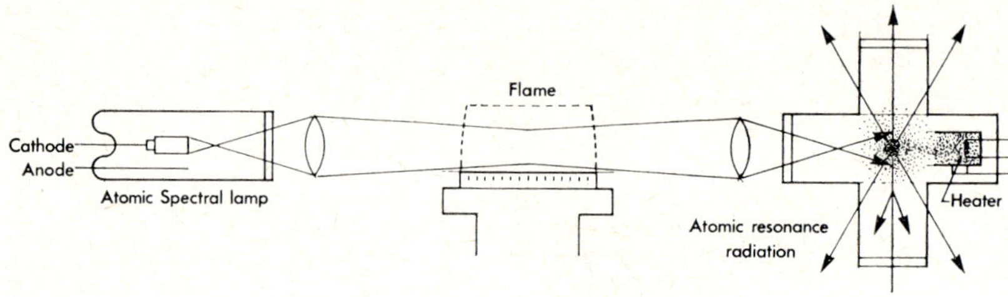

Fig. 7.7 The resonance detector

from the source, is collected through a side arm of the resonance lamp, and focused upon the photodetector. Since the radiation from the high intensity source is modulated, and that from the resonance lamp is unmodulated, only the re-emitted resonance radiation from the latter produces an output signal.

The resonance detector is thus not a true monochromator, since if the element being determined possesses more than one resonance line in its spectrum, energy from all these lines will be transmitted to the photodetector, and the sensitivity will be the average of the absorption sensitivities for all the lines. This will obviously be lower than that for the most absorbing line alone. The fact that the analytical sensitivity depends upon the spectral complexity of a particular element constitutes the major disadvantage of this technique. Other disadvantages are that resonance monochromators are difficult to stabilize, expensive to construct and of limited life expectancy.

Some advantages that have led to the manufacture of commercial resonance instruments and may well encourage their development for multi-channel on-line equipment are listed below:

(a) Resonance detectors cannot be put out of adjustment, in the same way that an optical monochromator can, by temperature changes and mechanical vibration. They could, therefore, be suitable for use under very rigorous conditions.

(b) No adjustment is needed to tune to a given line, so that their operation is even simpler than that of a standard atomic absorption spectrophotometer. They could be ideal for example, where very large numbers of samples are being analysed for a single element, especially if an automatic system were being employed.

(c) Flame noise is lower than that of a normal atomic absorption spectrophotometer.

SELECTIVE MODULATION[43]

This is a second technique that has been explored in some detail by WALSH and his group. The principle is illustrated in Fig. 7.8. Unmodulated light from a hollow-cathode lamp is passed along the axis of a discharge tube, the open ended cathode of which is made from the same element as the source lamp. The resonance radiation from the source lamp is thus absorbed by the concentration of atomic vapour within the open ended cathode, whereas all other radiation passes through undiminished in intensity.

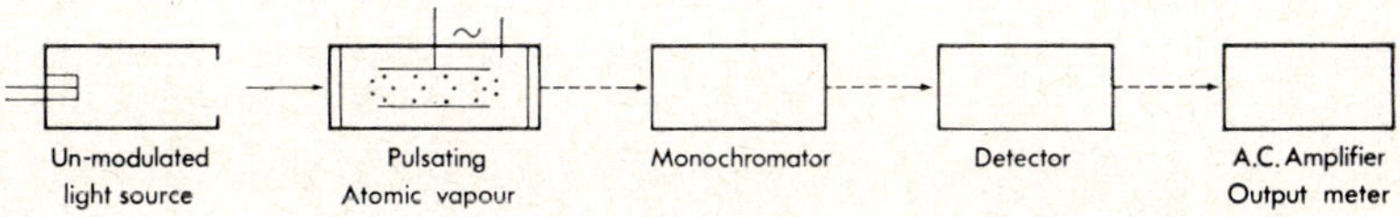

Fig. 7.8 Selective modulation of resonance lines

By modulating the power supply to the open ended cathode the atomic vapour concentration is made to pulsate, so that the resonance radiation from the source is *selectively modulated*, whereas all other radiation passes through the system unmodulated. An instrument employing this technique requires a monochromator, to eliminate extraneous radiation which is usually of sufficiently high intensity to cause unacceptable background noise.

WALSH has shown that selective modulation lamps can be built as one unit, with the hollow-cathode source and a modulating loop-electrode inside the same envelope. A gating circuit in the detection system removes any unwanted signals due to the modulation pulse.

The most important advantage of selective modulation derives from the fact that since the technique effectively isolates the required resonance line from all other radiation, absolutely linear calibration curves are obtained. Furthermore, a low resolution monochromator can be used, and stability of wavelength setting can be less stringent.

PULSED CURRENT OPERATION OF HOLLOW CATHODE LAMPS[44]

DAWSON and ELLIS increased the emission intensity from a conventional hollow-cathode lamp, by operating the device at a high current, passed through it in pulses of short duration. The lamp was run at a low d.c. current between pulses and the authors emphasized that this must be at a very low level. They claimed that there was no increase in resonance line width, and that self absorption did not occur. Also because the signal-to-noise ratio is drastically improved better limits of detection are obtainable.

Work in our own laboratory (E.E.L.) confirmed that the emission intensity from the lamp, and the signal-to-noise ratio, were considerably improved, but that the sensitivity was noticeably decreased. We felt that the most likely reason for this decrease in sensitivity was line broadening and/or self absorption, and were forced to conclude that the system was not satisfactory for inclusion into standard commercial equipment.

Atomic Fluorescence Spectroscopy (A F S)

Atomic Fluorescence Spectroscopy may be defined as the measurement of radiation from discrete atoms that are themselves being excited by the absorption of radiation from a given source which is not seen by the detector. Atomic fluorescence may be considered to be the analogue of molecular spectrofluorimetry and was first reported by WOOD[(45)] in 1904, when he succeeded in exciting fluorescence of the D lines of sodium vapour. This was achieved by illuminating sodium vapour contained in an evacuated test tube with light from a gas flame containing sodium chloride, and visually observing the yellow D lines.

WOOD called this fluorescence 'resonance radiation' because it was predicted by the classical theory of a light wave vibrating with the same frequency as the dipole oscillations of the medium. Soon after this initial discovery, resonance radiation was observed for mercury, cadmium, zinc and many other elements. This early work has been summarized by MITCHELL and ZEMANSKY[(46)] in their treatise Resonance Radiation and Excited Atoms.

The fluorescence of atoms in flames was first reported by NICHOLS and HOWES[(47)] in 1923 who obtained weak fluorescence from Ba, Ca, Li, Na, Sr, and Tl atoms present in high concentration in a hydrogen–air flame irradiated by light containing the resonance line of the appropriate metal.

Little work was done on atomic fluorescence spectroscopy after the 1930s until recently, when it has attracted new interest for two reasons. Firstly as a method of investigating the physical and chemical processes that occur in flames. Secondly as a basis of a new analytical technique theoretically possessing some advantages over both atomic absorption and flame emission methods for the detection and estimation of trace metallic elements.

In 1961 ROBINSON[48] observed weak fluorescence of the 2852 Å Mg line in an oxygen–hydrogen flame using a magnesium hollow-cathode lamp. The following year ALKEMADE[49] used the atomic fluorescence of sodium to study mechanisms of excitation and de-activation of atoms in flames and to measure the quantum efficiency for the 5890 Å sodium D line. He was the first to point out the possible analytical value of the technique.

The first analytical method was developed by WINEFORDNER and his co–workers in a series of four papers published in 1964–1965[50–53]. A year later WEST and his co-workers[54,55] at Imperial College used commercially available equipment for measuring atomic fluorescence and since then both research teams have greatly extended the application of this technique.

In atomic absorption spectrophotometry, the subsequent history of the energy absorbed by the atoms is of little concern. Much of the energy is lost by collisional deactivation within the flame gases. However, some of the energy imparted to the atoms is re-emitted in all directions and this phenomenon is the basis of atomic fluorescence.

TYPES OF ATOMIC FLUORESCENCE LINES

The basic atomic fluorescence arrangement, as shown in Fig. 7.9, consists of an intense source focused on to an atomic population in a flame or atom reservoir. Fluorescent radiation, which is emitted in all directions, is examined by a detector in the same plane at right angles to the incident light.

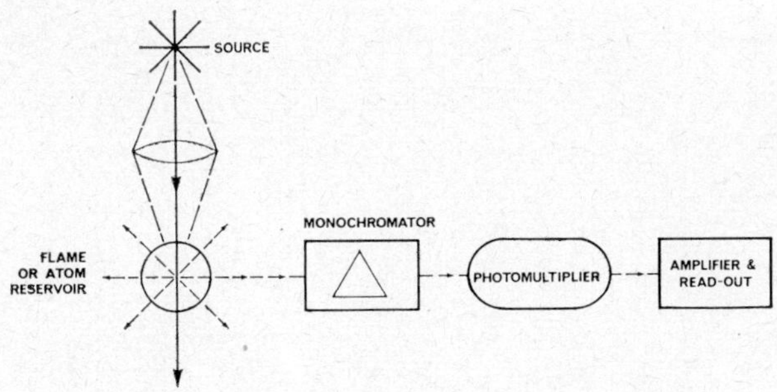

Fig. 7.9 Basic atomic fluorescent arrangement

The source may be either an atomic line source or a continuum and serves to excite atoms by the absorption of radiation of the proper wavelength. The atoms are then deactivated partly by collisional quenching with flame gas molecules and partly by emission of fluorescent radiation in all directions.

The wavelength of the fluorescent radiation is generally the same or longer than the incident radiation. The wavelength of the emitted radiation is characteristic of the absorbing atoms and the intensity of the emission may be used as a measure of their concentration.

This intensity is governed by the following relationship:

$$I_f = \phi I_0 \epsilon_A l k C$$

Where

I_f is the intensity of fluorescent radiation

C is the concentration of metal ion in solution

k is the proportionality constant to allow for efficiency of atomization and nebulization

ϵ_A is the atomic extinction coefficient of the atomic species

l is the length of flame

ϕ is the quantum efficiency for the fluorescent process.

ϕ may be defined as the ratio of the number of atoms which fluoresce from the excited state to the number of atoms which undergo excitation to the same excited state from the ground state in unit time.

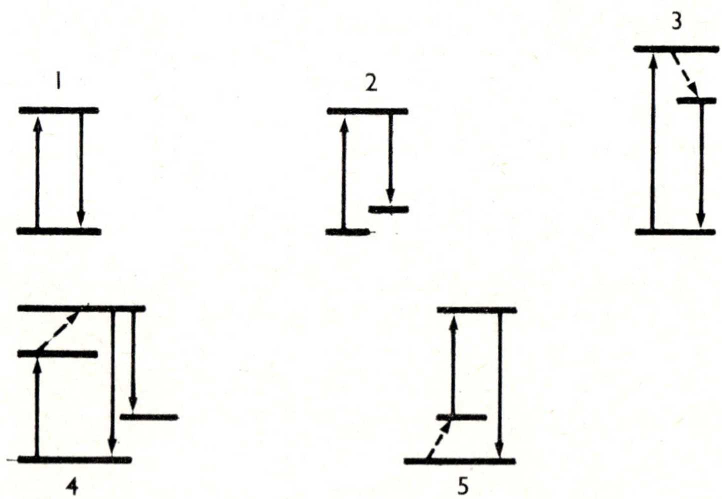

Fig. 7.10 Diagrammatic representation of the five basic types of fluorescence

There are five basic types of fluorescence which occur in flame measurement: (1) resonance fluorescence, (2) direct-line fluorescence,

(3) stepwise-line fluorescence, (4) thermally assisted direct-line fluorescence, (5) thermally assisted anti-Stokes fluorescence.

Resonance fluorescence. Resonance fluorescence occurs when an atom emits a spectral line of the same wavelength as used for excitation of the atom.

Direct-line Fluorescence. This occurs when an atom emits a spectral line of longer wavelength than the spectral line used for excitation (both the exciting and emitted lines have the same excited state). Thus after the fluorescence emission occurs the atom is in a metastable state above the ground state, e.g. emission of the 2386 Å tellurium line after excitation by the 2143 Å tellurium line.

Stepwise-line Fluorescence. This occurs when an atom emits resonance radiation after excitation of the atom to a higher energy state than the first resonance state followed by a radiationless deactivation to the first excited state, e.g. emission of the 3068 Å bismuth line after excitation the 2062 Å bismuth line[56].

Thermally assisted Direct-line Fluorescence. This type of fluorescence was first observed in a study of bismuth[56] atomic fluorescence characteristics and occurs when an atom is raised to an excited state by absorption of a resonance line and then is further excited to a slightly higher level by energy from the flame (i.e. collisions with flame gas molecules).

Fluorescence emission then occurs from this higher energy excited state either to the ground state or to a metastable state above the ground state, e.g. 2938 Å bismuth line after excitation by the 2062 Å bismuth line[56].

Thermally assisted Anti-Stokes Fluorescence. This type of process was reported for indium[57] and occurs when an atom is raised to a metastable state above the ground state by thermal excitation from the flame, then is excited by absorbing radiation to a higher energy excited state. Fluorescent emission may then occur from this excited state to the ground state. The wavelength of the fluorescent emission is then shorter than the wavelength of the exciting radiation, e.g. emission of the 4101 Å indium line after excitation by the 4511 Å indium line[57].

Combinations of the above processes therefore make up the final fluorescence spectrum observed when a flame containing an atomic population is excited either by an atomic line source or continuum.

Sensitized fluorescence has been observed in non-flame systems and occurs when an atom emits radiation after collisional activation by a foreign atom, which has been previously excited by the absorption of resonance radiation, e.g. excitation of thallium by mercury vapour excited by the 2537 Å mercury resonance line.

The following process takes place with the resulting sensitized fluorescence of the thallium 3776 and 5350 Å lines.

$$\mathrm{Hg}^* + \mathrm{Te} \rightarrow \mathrm{Te}^* + \mathrm{Hg}$$
$$\mathrm{Te}^* \rightarrow \mathrm{Te} + h\nu$$

* Excited atoms.

This type of fluorescence has been reported in non-flame cells[46].

The optimum conditions for the detection of metals in flames by atomic fluorescence has been analysed by JENKINS[64].

The advantages of atomic fluorescence have been discussed elsewhere. Current research is aiming to improve the sensitivity of determinations of trace metals. To do this many new modifications are being employed. The major improvement came with the use of microwave excited spectral sources[58]. These have been used by many workers to improve the sensitivity by increasing the intensity of the excitation source.

High intensity hollow-cathode lamps[59,60] and the carbon filament atom reservoir have also been successfully used as excitation sources. The sensitivity obtained by atomic fluorescence is generally better than by atomic absorption, although the interferences experienced in both techniques are similar.

Elements determined by atomic fluorescence in flames include As, Bi, Be, Ga, Ge, Hg, Mg, Sb, Se, Te, Tl, Zn. Many refractory elements can now be determined using the separated nitrous oxide–acetylene flame as an atom reservoir[63].

Chap. VII References

TECHNIQUES AND INSTRUMENTATION

1. LANG, R., Ultrasonic atomization of liquids. *J. Acoust. Soc. Amer.* 34 (1962) 6.
2. WENDT, R. H., and FASSEL, V. A., Induction-coupled plasma spectrometric excitation source. *Anal. Chem.* 37 (1965) 920.
3. WEST, C. D., and HUME, D., Radio-frequency plasma emission spectrophotometer *Anal. Chem.* 36 (1964) 412.
4. WEST, C. D., Ultrasonic sprayer for atomic emission and absorption Spectrochemistry. *Anal. Chem.* 40 (1968) 253.

5. Hoare, H. C., Mostyn, R. A., and Newland, B. T. N., An ultrasonic atomizer applied to A A S. *Anal. Chim. Acta* 40 (1968) 181.

6. Van Rensberg, H. C., and Zeeman, P. B., The determination of Au, Pt, Pd, Rh by A A S with an ultrasonic nebulizer and a multi-element high-intensity hollow cathode lamp with selective modulation. *Anal. Chim. Acta* 43 (1968) 173.

7. Kirsten, W. J., and Bertilsson, G. O. B., Direct continuous quantitative ultrasonic nebulizer for flame photometry and flame absorption spectrophotometry. *Anal. Chem.* 38 (1966) 648.

Flame Systems

8. Gatehouse, B. M., and Willis, J. B., Performance of a simple atomic absorption spectrophotometer. *Spectrochim. Acta* 17 (1961) 710.

9. Amos, M. D., and Willis, J. B., Use of high temperature premixed flames in atomic absorption spectroscopy. *Spectrochim. Acta* 22 (1966) 1325.

10. Foster, W. H., and Hume, D. N., Factors affecting emission intensities in flame photometry. *Anal. Chem.* 31 (1959) 2028.

11. Robinson, J. W., Recent advances in A A S. *Anal. Chem.* 33 (1961) 1067.

12. Robinson, J. W., Flame photometry using the oxy–cyanogen flame. *Anal. Chem.* 33 (1961) 1226.

13. Baker, M. R., and Vallee, B. L., A theory of spectral excitation in flames as a function of sample flow. *Anal. Chem.* 31 (1959) 2036.

14. Fassel, V. A., and Mossotti, V. G., Atomic absorption spectra of vanadium, titanium, niobium, scandium, yttrium, and rhenium. *Anal. Chem.* 35 (1963) 252.

15. Slavin, W., and Manning, D. C., Atomic absorption spectrophotometry in strongly reducing oxy-acetylene flames. *Anal. Chem.* 35 (1963) 253.

16. Skogerboe, R. K., and Woodriff, R. A., Atomic absorption spectra of europium, thulium and ytterbium using a flame as line sources. *Anal. Chem.* 35 (1963) 1977.

17. Amos, M. D., and Thomas, P. E., The determination of aluminium in aqueous solution by atomic absorption spectroscopy. *Anal. Chim. Acta* 32 (1965) 139.

18. WILLIS, J. B., A A S with high temperature flames. *Applied Optics* 7 (1968) 1295.

19. THOMPSON, K. C., Imperial College of Science and Technology, London. Unpublished Studies.

20. SLAVIN, W., VENGHIATTIS, A., and MANNING, D. C., Some recent experience with the nitrous oxide–acetylene flame. *At. Abs. Newsletter* 5 (1966) 84.

21. SMITHELLS, A., and INGLE, H., The structure and chemistry of flames. *J. Chem. Soc.* (*transactions*) 61 (1892) 204.

22. KIRKBRIGHT, G. F., SEMB, A., and WEST, T. S., Spectroscopy in separated flames—I The use of the separated air–acetylene flame in thermal emission spectroscopy. *Talanta* 14 (1967) 1011.

23. KIRKBRIGHT, G. F., SEMB, A., and WEST, T. S., The separated nitrous oxide–acetylene flame as an atom reservoir in thermal emission spectroscopy. *Spectroscopy Letters* 1 (1968) 7.

24. HINGLE, D., KIRKBRIGHT, G. F., and WEST, T. S., Spectroscopy in separated flames—II The use of the separated air–acetylene flame in long path absorption spectroscopy. *Talanta* 15 (1968) 199.

25. KIRKBRIGHT, G. F., SEMB, A., and WEST, T. S., Spectroscopy in separated flames—III Use of the separated nitrous oxide–acetylene flame in thermal emission spectroscopy. *Talanta* 15 (1968) 441.

26. HOBBS, R. S., KIRKBRIGHT, G. F., SARGENT, M., and WEST, T. S., Spectroscopy in separated flames—IV Application of the nitrogen-separated air–acetylene flame in flame-emission and atomic-fluorescence spectroscopy. *Talanta* 15 (1968) 997.

27. HINGLE, D. N., KIRKBRIGHT, G. F., and WEST, T. S., The determination of beryllium by thermal-emission and atomic fluorescence spectroscopy in a separated nitrous oxide–acetylene flame. *Analyst* 93 (1968) 522.

MODIFIED BURNER SYSTEMS

28. FUWA, K., and VALLEE, B. L., The physical basis of analytical atomic absorption spectrometry. *Anal. Chem.* 35 (1963) 942.

29. RUBEŠKA, I., and STUPAR, J., The use of absorption tubes for the determination of noble metals in atomic absorption spectroscopy. *At. Abs. Newsletter* 5 (1966) 69.

30. RUBEŠKA, I., and MOLDAN, B., Investigations on long-path absorption tubes in atomic absorption spectroscopy. *Analyst* 93 (1968) 148.

31. SPRAGUE, S., and SLAVIN, W., Performance of a three-slot Boling burner. *At. Abs. Newsletter* 4 (1965) 293.

32. BOLING, E. A., A multiple slit burner for atomic absorption spectroscopy. *Spectrochim. Acta* 22 (1966) 425.

33. ROBINSON, J. W., and HARRIS, B. M., Mechanical feed burner with total consumption for flame photometry and Atomic Absorption Spectroscopy. *Anal. Chim. Acta* 26 (1962) 439.

34. RAWSON, R. A. G., Improvement in performance of a simple atomic absorptionmeter by using pre-heated air and town gas. *Analyst* 91 (1966) 630.

NON-FLAME SAMPLING

35. L'VOV, B. V., The analytical use of atomic absorption spectra. *Spectrochim. Acta* 17 (1961) 761.

36. GATEHOUSE, B. M., and WALSH, A., Analysis of metallic samples by atomic absorption spectroscopy. *Spectrochim. Acta* 16 (1960) 602.

37. GOLEB, J. A., Uranium isotope investigations by atomic absorption. *Anal. Chem.* 35 (1963) 1978.

38. VENGHIATTIS, A. A., A technique for the direct sampling of solids without prior dissolution. *At. Abs. Newsletter* 6 (1967) 19.

39. WEST, T. S., and WILLIAMS, X. K., A A & A F S with a carbon filament atom reservoir. *Anal. Chim. Acta* 45 (1969) 27.

40. L'VOV, B. V., The potentialities of the graphite crucible in atomic absorption spectroscopy. *Spectrochim. Acta* 24 (1969) 53.

MODIFIED ATOMIC ABSORPTION SYSTEMS

41. SULLIVAN, J. V., and WALSH, A., Resonance radiation from atomic vapours. *Spectrochim. Acta* 21 (1965) 727.

42. SULLIVAN, J. V., and WALSH, A., The application of resonance lamps as monochromators in atomic absorption spectroscopy. *Spectrochim. Acta* 22 (1966) 1843.

43. BOWMAN, JUDITH A., SULLIVAN, J. V., and WALSH, A., Isolation of atomic resonance lines by selective modulation. *Spectrochim. Acta* 22 (1961) 205.

44. DAWSON, J. B., and ELLIS, D. J., Pulsed current operation of hollow cathode lamps to increase the intensity of resonance lines for atomic absorption spectroscopy. *Spectrochim. Acta* 23A (1967) 565.

ATOMIC FLUORESCENCE SPECTROSCOPY

45. WOOD, R. W., The fluorescence of sodium vapour and the resonance radiation of electrons. *Phil. Mag.* 10 (1905) 513.

46. MITCHELL, A. C. G., and ZEMANSKY, M. W., *Resonance Radiation and Excited Atoms.* University Press, Cambridge. 1961.

47. NICHOLS, E. L., and HOWES, H. L., The photo luminescence of flames. *Phys. Rev.* 22 (1923) 425; 23 (1924) 472.

48. ROBINSON, J. W., Mechanism of elemental spectral excitation in flame photometry. *Anal. Chim. Acta* 24 (1961) 254.

49. ALKEMADE, C. Th. J., *Proc. Xth Colloquium Spectroscopium Internationale* 1962 p. 143. Spartan Books, Washington D.C., 1963.

50. WINEFORDNER, J. D., and VICKERS, T. S., Atomic fluorescence spectrometry as a means of chemical analysis. *Anal. Chem.* 36 (1964) 161.

51. WINEFORDNER, J. D., STAAB, R. A., Determination of zinc, cadmium and mercury by atomic fluorescence flame spectrometry. *Anal. Chem.* 36 (1964) 165.

52. WINEFORDNER, J. D., and STAAB, R. A., Study of experimental parameters in atomic fluorescence flame spectrometry. *Anal. Chem.* 36 (1964) 1367.

53. WINEFORDNER, J. D., MANSFIELD, J. M., and VEILLON, C., High sensitivity determination of zinc, cadmium, mercury, thallium, gallium and indium by atomic fluorescence flame spectrometry. *Anal. Chem.* 37 (1965) 1049.

54. DAGNALL, R. M., WEST, T. S., and YOUNG, P., Determination of cadmium by atomic fluorescence spectroscopy and atomic absorption spectroscopy. *Talanta* 13 (1966) 803.

55. DAGNALL, R. M., THOMPSON, K. C., and WEST, T. S., An investigation of some experimental parameters in atomic fluorescence spectroscopy. *Anal. Chim. Acta* 36, (1966) 269.

56. DAGNALL, R. M., THOMPSON, K. C., and WEST, T. S., Studies in atomic fluorescence spectroscopy—VI. The fluorescence characteristics and analytical determination of bismuth with an iodine electrodeless discharge tube as source. *Talanta* 14 (1967) 1151.

57. OMENETTO, N., and ROSSI, G., Some observations on direct line fluorescence of thallium, indium, gallium. *Spectrochim. Acta* 24B (1969) 95.

58. DAGNALL, R. M., THOMPSON, K. C., and WEST, T. S., Microwave-excited electrodeless discharge tubes as spectral sources for atomic fluorescence and atomic-absorption spectroscopy. *Talanta* 14 (1967) 551.

59. WEST, T. S., and WILLIAMS, X. K., Atomic fluorescence spectroscopy of silver using a high intensity hollow cathode lamp as source. *Anal. Chem.* 40 (1968) 335.

60. WEST, T. S., and WILLIAMS, X. K., Atomic fluorescence spectroscopy of magnesium with a high intensity hollow cathode lamp as line source. *Anal. Chim. Acta* 42, (1968) 29.

61. MASSMAN, H., Studies of atomic absorption and atomic fluorescence spectroscopy in graphite cell. *Spectrochim. Acta* 23B (1968) 215.

62. WEST, T. S., and WILLIAMS, X. K., Atomic absorption and atomic fluorescence spectroscopy with a carbon filament atom reservoir. *Anal. Chim. Acta* 45 (1969) 27.

63. HINGLE, D., KIRKBRIGHT, G. F., and WEST, T. S., The determination of beryllium by thermal emission and atomic fluorescence spectroscopy in a separated nitrous oxide acetylene flame. *Analyst* 93 (1968) 522.

64. JENKINS, D. R., An analysis of the optimum conditions for the detection of metals in flames by atomic fluorescence. *Spectrochim. Acta* 23B (1968) 47.

VIII

Theory

By K. C. THOMPSON, Ph.D., B.Sc., D.I.C., A.R.C.S.

History

THE phenomenon of atomic absorption was first noticed by WOLLASTON[1] in 1802, when he observed a few dark lines in the solar spectrum. Then in 1814, FRAUNHOFER[2] observed the numerous dark lines in the solar spectrum that now bear his name. Although FRAUNHOFER could not explain their origin he made a map of about 700 lines and assigned the letters A to H to the eight most prominent ones. (e.g. the sodium D lines).

The basic principles underlying atomic absorption were established by KIRCHHOFF[3] in 1860, when he put forward the general law relating to the absorption and emission of light from a given system. KIRCHHOFF showed that a flame containing sodium chloride would not only emit the yellow sodium D lines, but also absorb the same yellow light from a continuous source placed behind the flame, no other wavelengths being absorbed. Thus the FRAUNHOFER lines were attributed to the absorption by certain elements, in the outer cooler solar atmosphere, of the continuous spectrum emitted by the hot interior of the sun.

KIRCHHOFF was the first person to emphasize the great significance of the characteristic spectra of the different elements. This important observation was the foundation of analytical spectroscopy.

The theory of atomic spectroscopy was evolved in the early twentieth century by physicists and astrophysicists. Much of the work of these early researchers (summarized in a treatise by MITCHELL and ZEMANSKY[4]) was performed at low pressures in enclosed vessels, and was not directed towards analytical purposes, except for some astrophysical studies on the determination of the compositions of the solar and stellar atmospheres. A special case was the estimation of the contamination of air by mercury, which has an appreciable vapour pressure at room temperature.

Although emission methods of analysis (arc, spark, and flame) became firmly established, it was not until 1953 that WALSH[5] realized the analytical potential of atomic absorption and demonstrated the superiority of the technique over flame emission spectroscopy. The first commercial instruments appeared about 1960, and since that time there has been an exponential increase in the number of atomic absorption papers published.

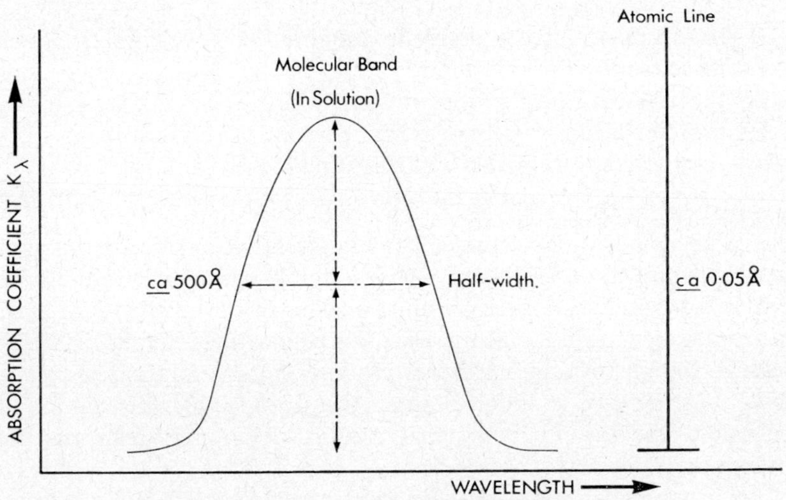

Fig. 8.1 Typical molecular band (in solution)

Absorption and Emission Line Profiles

A typical absorption line profile of an atomic vapour in a conventional flame and a typical profile of a molecular absorption band in solution are shown in Fig. 8.1.

The ordinate represents the absorption coefficient (K_λ) and the half width is the width across the profile where the absorption coefficient is half its maximum value. The absorption coefficient for a path length L of uniform atomic vapour (Fig. 8.2) is defined by the equation:

$$I_{T\lambda} = I_{0\lambda} \exp(-K_\lambda L)$$

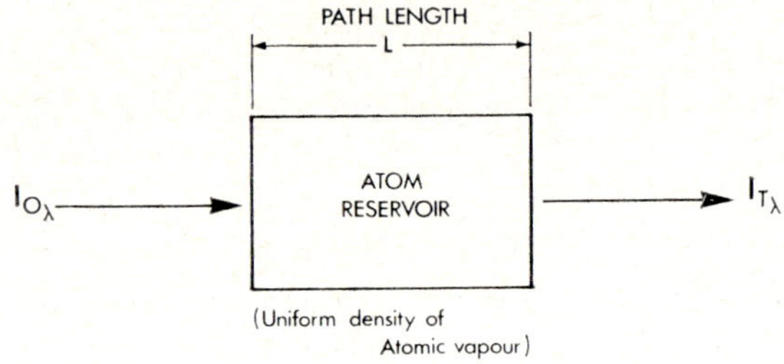

Fig. 8.2 Pictorial representation of atomic absorption

where

$I_{0\lambda}$ = the intensity of the incident radiation at wavelength λ per unit wavelength with no atomic vapour in the beam.
$I_{T\lambda}$ = the intensity of the transmitted radiation at wavelength λ per unit wavelength with an atomic vapour in the beam.
L = the path length of the uniform vapour (cm)
K_λ = the absorption coefficient (cm^{-1}).

The large half-widths of molecular electronic absorption bands allow the absorption to be measured using a continuous source and an inexpensive low resolution monochromator. The small half-width of atomic absorption line profiles means that if a continuous source is used to measure the absorption, an expensive very high resolution monochromator must be used in order to obtain any appreciable absorption. This is possibly one reason why the analytical uses of atomic absorption were not exploited earlier.

The large half-widths of molecular electronic absorption bands in solution (100–500 Å) can be ascribed to vibrational and rotational fine structure of the electronic energy levels and also to solvent solute interaction in the condensed phase. This is one reason why spectral interference due to overlapping of absorption line profiles is common in solution spectrophotometry.

The half-widths of absorption and emission lines is of considerable importance in atomic absorption spectroscopy and a summary of the broadening processes involved is given below.

Broadening Processes of Atomic Spectral Lines

For a given system in thermal equilibrium the profile of a resonance emission line is the same as the profile of the same resonance line in absorption (i.e. they have the same half-width). If a solution containing sodium atoms is nebulized into a flame the half-width of the D_1 emission line will be the same as the half-width of the D_1 absorption line. The main types of broadening processes of atomic lines are:

Natural Broadening

Natural broadening is due to the finite lifetime of the atom in the excited state (i.e. Heisenberg's uncertainty principle) and is independent of the environment of the atom. For most resonance lines the natural width is of the order of 10^{-4} Å which is negligible compared to that due to other causes in analytically used flames or sources.

Doppler Broadening

Doppler broadening is caused by the absorbing or emitting atoms having different component velocities along the line of observation. The broadening is symmetrical about the mean wavelength of the line. The Doppler half-width is proportional to the square root of the absolute temperature, independent of the pressure, and is given by the equation:

$$\Delta\lambda_D = \frac{\lambda_0}{c}\sqrt{\frac{8(\ln 2)RT}{M}} \qquad (1)$$

where

$\Delta\lambda_D$ = the Doppler half-width (Å)
λ_0 = the wavelength of the centre of the line (Å)
c = the velocity of light (cm s^{-1})
R = the gas constant (ergs °K^{-1} mole^{-1})
M = the atomic weight of the absorber (g mole^{-1})
T = the temperature of the atoms (°K).

For an element of atomic weight 75 in a flame at 2500°K the Doppler half-width varies from 0·008 Å at 2000 Å to 0·032 Å at 8000 Å. Therefore lines in the visible region have a greater Doppler half-width than lines in the ultra-violet. Also at a given wavelength the smaller the atomic weight of the element, the greater the Doppler half-width.

COLLISIONAL BROADENING (ALSO KNOWN AS PRESSURE OR LORENTZ BROADENING)

Collisional broadening is due to the perturbation of the absorbing or emitting atoms by foreign gas atoms. The effect of this type of broadening varies for different foreign gases and for different atomic states. Assuming the broadening is symmetrical, the half width is given by the equation:[6]

$$\Delta\lambda_c = \frac{2\lambda_0'^2\sigma_c^2 P_f}{\pi c k T}\left[2\pi R T\left\{\frac{1}{M_a}+\frac{1}{M_f}\right\}\right]^{1/2} \qquad (2)$$

where:

$\Delta\lambda_c'$ = the collisional half-width (cm)
λ_0' = the wavelength of the centre of the line (cm)
σ_c = the effective cross section for collisional broadening for a given line under given conditions (cm)
P_f = the pressure of the foreign gases (dynes cm^{-2})
M_a = the atomic weight of the absorbing or the emitting species (g $mole^{-1}$)
M_f = the effective molecular weight of the foreign gas molecules (g $mole^{-1}$)
R = the gas constant (erg $°K^{-1}$ $mole^{-1}$)
T = the temperature of the foreign gases (°K)
c = the velocity of light (cm sec^{-1})
k = the Boltzmann constant (erg $°K^{-1}$).

Thus

$$\Delta\lambda_c(\text{Å}) = 10^8 \times \Delta\lambda_c' \text{ (cm)}.$$

It can be seen, therefore, that the collisional half-width is proportional to the pressure, and inversely proportional to the square root of the temperature of the emitting or absorbing atoms. The half-width is also proportional to the square of the wavelength of the line. Owing to an uncertainty in the values of the collisional cross sections, the calculation of collisional half-widths is difficult and the results are not always consistent[6,7,8]. The effect of collisional broadening is to enhance the wings of the line compared to Doppler broadening (see Fig. 8.3). It has been stated that collisional broadening can also cause the line profile to become asymmetric[4] (Fig. 8.3). Thus the centre of a line in a high-pressure source can be at a very slightly different wavelength from the centre of the same line in a low pressure source. The wavelength shift in flames, at atmospheric pressure, compared to low pressure sources is very small[9].

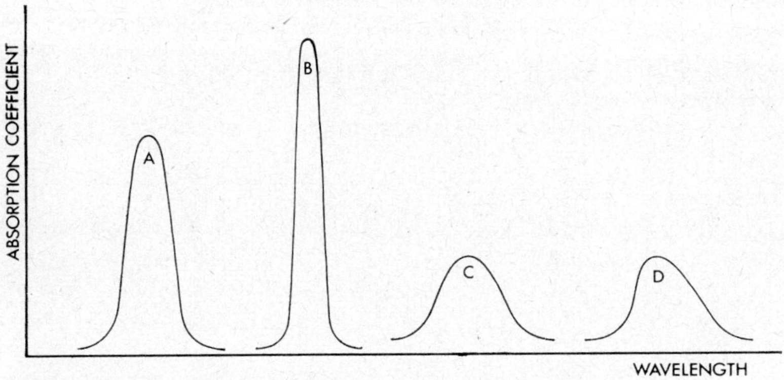

A. Typical Absorption Line Profile in a flame.(Doppler and Collisional Broadening.)
B. Doppler Broadened Absorption Line.(Negligible Collisional Broadening.)
C. Collision Broadened Absorption Line.(Negligible Doppler Broadening.
D. Collision Broadened Absorption Line at very high pressure (Ref.4, P.159) showing Asymmetry (exaggerated)

Fig. 8.3 Typical profiles of absorption lines

Resonance Broadening

Resonance broadening is due to the perturbation of the absorbing or emitting atom by atoms of the same type. In conventional flames the partial pressure of metal atoms, even at very high concentrations, seldom exceeds 10^{-3} Torr, thus this type of broadening process can be considered negligible in flames at 760 Torr[10].

Stark and Zeeman Broadening

Stark and Zeeman broadening are caused by electric and magnetic fields respectively. Although Stark broadening can cause lines to become diffuse in the arc and spark, both these effects are negligible in flames. It is just feasible that the Stark effect could very slightly influence the lines from a hollow-cathode lamp, but at the voltages and currents employed in most lamps this effect is thought to be negligible.

Hyperfine Structure

Hyperfine structure, which is not an actual broadening process, can be attributed to a non-zero value of the nuclear spin and/or the presence of several isotopes[4,11,12]. Thus each line consists of a number of separate hyperfine components each acting as an independent line. For all isotopes with an even atomic number and mass the nuclear spin is zero. Thus for an element with an even atomic number and just one

TABLE VIII-1. HYPERFINE SPLITTING DUE TO NUCLEAR SPIN

Element	Naturally occurring isotopes	Natural % abundance	Spin of nucleus	Line (Å)
Hg	198	10·0	0	any
	199	16·8	1/2	2537
	200	23·1	0	any
	201	13·2	3/2	2537
	202	29·8	0	any
	204	6·8	0	any
Bi	209	100	9/2	3068
				2228
				2062
Th	232	100	0	any
Tl	203 }	29·5 }	1/2	3776
	205 }	70·5 }	1/2	
Li	6	7·4	1	6708
	7	92·6	3/2	
Na	23	100	3/2	5890
K	39	93·1	3/2	7665
	41	6·9	3/2	
Rb	85	72·2	5/2	7800
	87	27·8	3/2	7800
Cs	133	100	7/2	8521
Au	197	100	3/2	2428
U	235	0·72	7/2	4153
				5027
	238	99·3	0	any
Pb	204	1·4	0	any
	206	25·1	0	any
	207	21·2	1/2	2833
	208	52·3	0	any
Mn	55	100	5/2	4033

[a] Extreme hyperfine splitting is the sum of the splitting of the upper state and the splitting of the lower state.

Extreme[a] hyperfine splitting (Å)	Number of components		References	Element
	Upper energy level	Lower energy level		
0	1	1	4, 17	Hg
0·047	2	1		
0	1	1		
0·047	3	1		
0	1	1		
0	1	1		
>0·078[b]	2	4	11	Bi
>0·070[b]	4	4	11	
	6	4		
0	1	1		Th
0·16	2	2	11	Tl
0·012[c]	Splitting of the upper $^2P_{1\frac{1}{2}}$ state, at least an order of magnitude smaller than that of ground state $^2S_{\frac{1}{2}}$.	2	12	Li
0·020[c]		2	4, 12	Na
0·009[c]		2	12	K
		2	12	
0·061[c]		2	18, 12	Rb
0·14[c]		2	18	
0·22[c]		2	12, 19	Cs
0·012[c]		2	12	Au
smaller than isotopic shift	—	—	20	U
0	1	1		
0	1	1		Pb
0	1	1	46	
0·036	2	1		
0	1	1		
0·042	6	6	29	Mn

[b] Represents splitting of the upper level only, no results for the splitting of the ground $^4S_{1\frac{1}{2}}$ state could be found.

[c] Represents splitting of lower $^2S_{\frac{1}{2}}$ level only. The splitting of the upper ^2P levels is at least one order of magnitude smaller[12].

even atomic mass isotope (e.g. naturally occurring thorium exists solely as Th^{232}), there is no hyperfine structure to the spectrum. This is why lamps containing an element with an even atomic number and a single even atomic mass isotope are used as wavelength standards (Kr^{86}, Hg^{198}, Th^{232}) [13,14,15]

The separation of the hyperfine components due to nuclear spin, and different isotopes, is small and difficult to measure, and it is difficult to evaluate the total hyperfine separation compared to the total absorption line half-width in conventional flames. Although it has often been stated that hyperfine splitting is negligible compared to absorption line half-widths, this cannot be considered to be true for all elements because isotopic analysis of lithium 6 and 7 in flames[16] by atomic absorption has been reported, and MITCHELL and ZEMANSKY state that the 2537 Å mercury line, from naturally occurring mercury, is split into 9 components of which 5 are resolvable using an interferometer. The extreme hyperfine splitting was given as 0·047 Å which is comparable to the absorption line width in conventional flames[6]. Other values are quoted in Tables VIII-1 and VIII-2.

Each hyperfine component will be broadened by the processes described above.

Total Line Half-widths. In conventional flames the two most important broadening processes (hyperfine structure not being an actual broadening process) are Doppler and Collisional broadening. The total line half-width of each hyperfine component is given by:

$$\Delta\lambda_T = [(\Delta\lambda_D)^2 + (\Delta\lambda_c)^2]^{1/2} \qquad (3)$$

where

$\Delta\lambda_T$ = the total line half-width of each hyperfine component (Å)
$\Delta\lambda_D$ = the half-width due to Doppler broadening alone (Å)
$\Delta\lambda_c$ = the half-width due to collisional broadening alone (Å)
$a = \dfrac{\Delta\lambda_c \,(\ln 2)^{1/2}}{\Delta\lambda_D}$

a is called the damping constant and for most absorption lines in conventional flames is thought to have a value of between 0·5 and 2. Thus $\Delta\lambda_D$ and $\Delta\lambda_c$ are of the same order of magnitude.

The measurement of the half-widths of absorption line profiles is a complicated process, and the results obtained are not always consistent, owing to an uncertainty in the values of the collisional cross sections involved. SOBOLEV[24] has determined the total half-widths of the main resonance lines of calcium (4227 Å), lithium (6708 Å), and sodium

TABLE VIII-2. HYPERFINE SPLITTING DUE TO NATURALLY OCCURRING ISOTOPES

Element	Naturally occurring isotopes	Natural % abundance	Spin of nucleus	Isotope shift (Å)	Isotopes shift measured between	Line (Å)	References
Hg	198	10·0	0	0·010	198–200	2537	4, 17
	199	16·8	1/2	0·012	200–202		
	200	23·1	0	0·010	202–204		
	201	13·2	3/2	0·032	198–204		
	202	29·8	0				
	204	6·8	0				
Bi	209	100	9/2	0		any	
Th	232	100	0	0		any	15
Tl	203	29·5	1/2	0·016	203–205	5350	11
	205	70·5	1/2				
Li	6	7·4	1	0·16	6–7	6708*	21, 22
	7	92·6	3/2				
B	10	18·7	3	0·01	10–11	2497, 2498	23
	11	81·3	3/2				
Na	23	100	3/2	0		any	
Rb	85	72·2	5/2	Less than hyperfine splitting of 87 Rb		7800	18
	87	27·8	3/2				
U	235	0·72	7/2	0·07	235–238	4153	20
	238	99·3	0	0·10	235–238	5027	20
Cs	133	100	7/2	0		any	
Pb	204	1·4	0				
	206	25·1	0	0·007	206–208	2833	46
	207	21·2	½	0·021	206–207a**		
	208	52·3	0	0·008	207b–208**		

* Doublet (see Fig. 8.12).
** See Fig. 8.13.

(5890 Å) in the air–acetylene flame, obtaining values of $9{\cdot}1\times10^{-2}$, $1{\cdot}3\times10^{-1}$ and 5×10^{-2} Å respectively.

WINEFORDNER, PARSONS and MCCARTHY[6] have theoretically calculated the maximum and minimum values of the total line half-widths of 60 elements in various types of flame (hyperfine splitting was ignored). The Doppler half-widths were calculated by substitution of the known flame temperatures and the atomic weights of the elements in equation (1). The collisional half-widths were calculated, using the same flame temperatures, an effective molecular weight of the flame gases, and assuming a minimum collisional cross section of 20Å^2 in conjunction with equation (2). This gave a minimum value of the collisional half-width. By assuming a maximum collisional cross section of 100 Å^2, a maximum value of the collisional half-width was obtained. Then using equation (3) the maximum and minimum total half-widths were calculated. For most elements the average total line half-width lies in the range $10^{-1} - 10^{-2}$ Å, the total half-width tending to increase with increasing wavelength. This is because the Doppler half-width varies directly with wavelength, and the collisional half-width varies with the square of the wavelength.

The effect of hyperfine structure will be to increase the effective line half-width values for many elements. This is borne out by studies of the overlap of emission line with absorption line profiles. For instance, the 2061·63 Å iodine line (emitted by a heavily cooled iodine electrodeless discharge tube) is fairly strongly absorbed by bismuth atoms[25,26] in air–propane and air–acetylene flames. The corresponding bismuth absorption line is at 2061·70 Å, a wavelength difference of 0·07 Å. Assuming the effective Doppler temperature of the electrodeless discharge tube to be about 1000°K, and neglecting collisional broadening, the total line half-width of the 2061·63 Å iodine line is 0·004 Å. Using WINEFORDNER and PARSONS' method[6], the maximum bismuth 2061·70 Å absorption line half-width in the air–acetylene flame is 0·007 Å. Thus it would appear that the line profiles must have a somewhat greater half-width than the maximum values calculated ignoring hyperfine structure. This is further substantiated by the fact that the 2288·12 Å arsenic line from an electrodeless discharge tube is weakly absorbed by cadmium atoms in the air–acetylene flame[25,26]. The corresponding cadmium absorption line is at 2288·02 Å, a wavelength difference of 0·1 Å. (The arsenic lamp did not emit a cadmium spectrum, and this means that the absorption was due to overlap of the arsenic and cadmium line profiles.) Assuming a Doppler temperature of the source of 1000°K and neglecting collisional broadening, the total line half-width of the arsenic 2288·12 Å line is 0·006 Å. Using WINEFORDNER and PARSONS' method, the maximum cadmium 2288·02 Å absorption line half width in the air–acetylene flame is 0·011 Å. Similarly ALLAN[29] has stated that hyperfine structure is the major factor involved in the overlap of the manganese 4033·07 Å and gallium 4032·98 Å lines.

It is important to realize that the total effective width of an absorption line is not necessarily twice the half-width. In fact the wings of a line can extend some way beyond this, but the above overlaps are thought to be more likely due to the hyperfine structure than absorption by the wings of the absorption line. (For bismuth see Table VIII-1).

The narrow half-width of absorption line profiles means that a continuous source cannot be used to measure the absorption unless a very high resolution monochromator is used ($<0{\cdot}2$Å). WALSH[5] overcame this problem by using an atomic line source (hollow-cathode lamp) that emitted the atomic line spectrum of the element to be determined. The low pressure in hollow-cathode lamps means that there is negligible collisional broadening, also the Doppler temperature of the discharge is somewhat less than that of a flame[9]. Thus the total half-width of a resonance emission line from a hollow-cathode lamp is expected to be less than the half-width of the absorption profile for the same line in a conventional flame. An inexpensive, low resolution monochromator can, therefore, be used to isolate the desired resonance line emitted by the hollow-cathode lamp, from any other unwanted lines.

The combination of an atomic line source and a low resolution monochromator is equivalent to a very high resolution monochromator in conjunction with a continuum source (see Fig. 8.4).

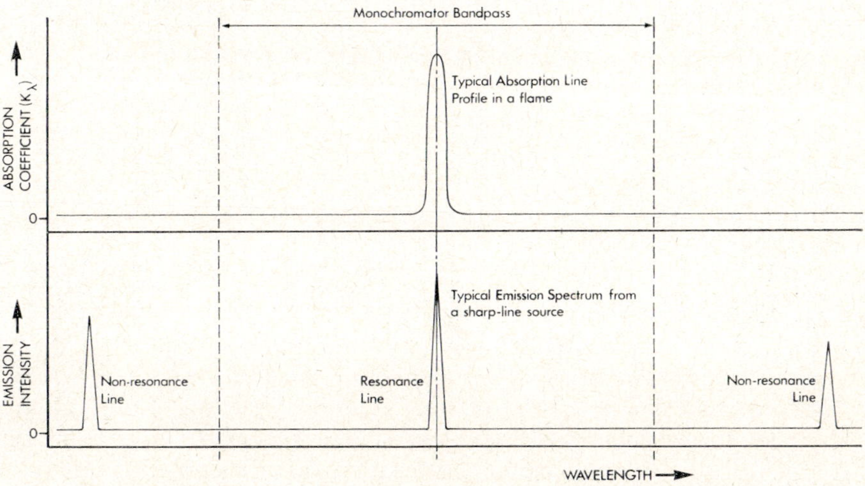

Fig. 8.4 Combination of a hollow-cathode lamp and a relatively low resolution monochromator

Spectral Interference in Atomic Absorption

The narrow width of absorption lines results in very little spectral interference in atomic absorption. The average line half-width is about 0·05 Å and the wavelength region over which usable lines occur is about 6000 Å (2000–8000 Å), thus there is only remote possibility of overlap of the limited number of absorption line profiles. In solution spectrophotometry with molecular band half-widths of about 100–500 Å the possibility of overlap is much greater.

Although spectral interference in atomic absorption is rare it has been observed. FASSEL, RAMUSON and COWLEY[27] have observed spectral interference between the following absorption lines: copper 3247·54 Å and europium 3247·53 Å ($\Delta\lambda = 0{\cdot}01$ Å), iron 2719·03 Å and platinum 2719·04 Å ($\Delta\lambda = 0{\cdot}01$ Å), silicon 2506·90 Å and vanadium 2506·91 Å ($\Delta\lambda = 0{\cdot}01$ Å), aluminium 3082·16 Å and vanadium 3082·11 Å ($\Delta\lambda = 0{\cdot}05$ Å). MANNING and FERNANDEZ[28] observed spectral interference between the mercury 2536·52 Å and cobalt 2536·49 Å lines ($\Delta\lambda = 0{\cdot}03$ Å).

ALLAN[29] has observed interference between the gallium 4032·98 Å and the manganese 4033·07 Å lines ($\Delta\lambda = 0{\cdot}09$ Å). Thus if it were necessary to determine copper in the presence of europium, the 3274 Å line should be used rather than the 3248 Å line. Also a correction would be necessary if mercury were to be determined in the presence of large amounts of cobalt.

All other examples quoted were for little-used absorption lines, hence it is true to say that spectral interference is seldom observed under normal working conditions in atomic absorption.

The Shape of Calibration Curves in Atomic Absorption

CASE 1

The absorption line half-width is greater than the source line half-width, but is less than the monochromator bandpass.

This is generally the case for hollow-cathode lamps and electrodeless discharge tubes. These sources operate at low pressures and should show very little collisional broadening. The main cause of broadening is due to the Doppler effect. YASUDA[9] has estimated the effective temperature of a calcium hollow-cathode lamp, operated at 50 mA to be about 1400°K. Most lamps are run at considerably lower currents and should have an even lower Doppler temperature. The actual operating temperature of most electrodeless discharge tubes is relatively low, and these should also exhibit a low Doppler temperature, and negligible collisional broadening. Thus the half-widths of resonance lines from these sources are expected to be less than the half-widths of absorption lines in flames, because there will be appreciable collisional broadening and a relatively high Doppler temperature in most conventional flames.

The effect of the hyperfine splitting of spectral lines, due to nuclear spin and different isotopes, is difficult to assess, as relatively little work has been performed on this aspect of atomic absorption. It is possible that the half-widths of resonance lines emitted by electrodeless discharge tubes (and also by hollow-cathode lamps) are sufficiently small for some of the hyperfine structure to be partially or completely resolved[18,19]. There is less chance that some of the hyperfine components of the wider absorption lines in flames are resolvable.

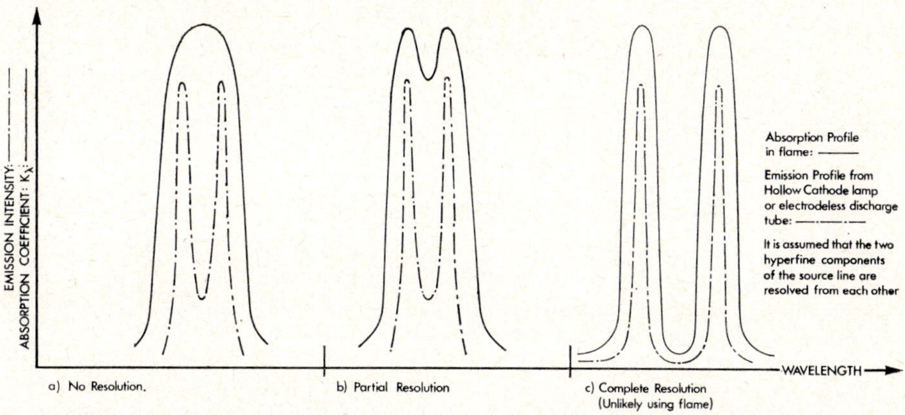

Fig. 8.5 Possible shapes of the absorption profile and the emission profile of a line with two hyperfine components

There are three cases to be considered, firstly, the hyperfine splitting will be small compared to the absorption line half-width, secondly, there might be partial resolution of some of the absorption line components, and lastly the main hyperfine components will be resolved. In Fig. 8.5 it is assumed that the two hyperfine components of the emission line are resolved.

It is thought that the last possibility (complete resolution) is very unlikely in conventional flames, although it has been reported in low-pressure absorption cells[17] (see isotopic analysis, later).

Assuming that the absorption coefficient (K_λ) is effectively constant across the emission line width(s), then the Beer Lambert Law can be applied (see Fig 8.6).

$$-\int_{I_T}^{I} \frac{dI}{I} = \int_{L}^{0} K\,N\,dl \tag{4}$$

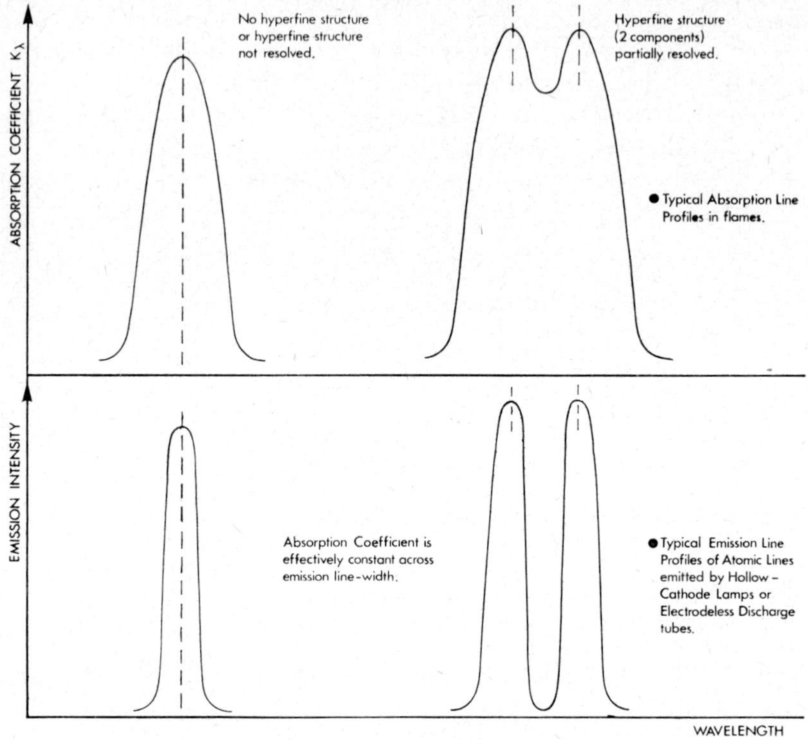

Fig. 8.6 Atomic absorption using a narrow line source

where

I_0 = the integrated intensity of the resonance line from the source passing through the flame and reaching the detector when nebulizing a blank solution (erg s^{-1} cm^{-2})

I_T = the integrated intensity of the source line after absorption by atoms in the flame when a sample solution is nebulized into the flame (erg s^{-1} cm^{-2})

L = the path length through the flame (cm)

N = the concentration of atoms in the flame (cm^{-3})

K = constant ($KN = \bar{K}_\lambda$) (cm^{-1})

$\bar{K}_\lambda$ = effective absorption coefficient across the emission line profile (cm^{-1})

$$\therefore \qquad \log_e \frac{I_0}{I_T} = KNL \qquad (5)$$

$$\therefore \qquad \text{Absorbance} = A = \log_{10} I_0/I_T = 0{\cdot}434KNL.$$

The absorbance ($\log_{10} I_0/I_T$) is proportional to the number of atoms in the flame. In general, for a resonance line from a sharp line source such as a hollow-cathode lamp or an electrodeless discharge tube (assuming that the resonance line is resolved from all other lines), the absorbance is directly proportional to concentration of the sample for absorbances less than about 0·5–1·0, and for absorbances greater than this the absorbance increases less than proportionally with concentration (see Fig. 8.8). This curving off of the calibration curve is probably due to the fact that at high absorbances the absorption coefficient cannot be considered to be effectively constant across the emission line width, or line widths in the case of hyperfine structure, (see Fig. 8.5). At high absorbances the shape of the wings of the line become important (Fig. 8.3). The curvature of the calibration curve is unlikely to be due to resonance broadening[(10)], as even when high concentrations of the test element are nebulized into the flame, the partial pressure of the test element will seldom exceed 10^{-3} Torr.

Case 2

The absorption line half-width is less than the source line half-width and is less than the monochromator bandpass.

This is the case for a continuous source, e.g. xenon arc lamp, deuterium lamp, quartz-iodine lamp, and the absorption is now critically dependent on the bandpass of the monochromator (Fig. 8.7). If a continuous source is used it is essential to employ a monochromator with a spectral bandpass $<0{\cdot}2$ Å, otherwise very poor limits of detection will be obtained. The spectral bandpass of most monochromators is greater than the total absorption line profile width (Fig. 8.7) and this fact is assumed for the derivation given below[(30)].

Assumptions

1. Constant temperature throughout the flame.
2. Constant and uniform atomic concentrations throughout the flame.
3. Constant transmission of light within monochromator spectral bandpass and zero transmission at all other wavelengths (Fig. 8.7).
4. The intensity of the continuum source is constant over the wavelength region of the absorption line profile.
5. Flame uniformly illuminated by a parallel beam of light.
6. The spectral bandpass of the monochromator is greater than the total absorption line profile width (Fig. 8.7).

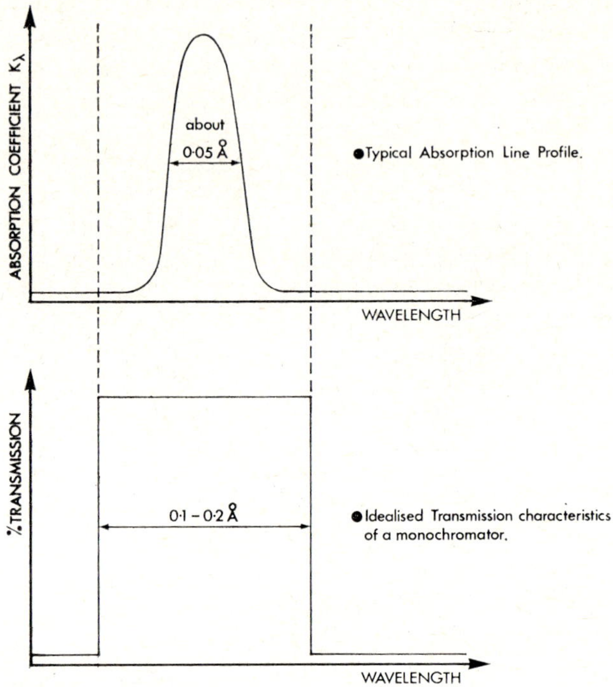

Fig. 8.7 Atomic absorption using a continuous source

Let

λ_0 = the peak wavelength of the absorption line profile (cm)

$I_{0\lambda}$ = the intensity of the continuum source at wavelength λ that is passing through the flame (erg sec^{-1} cm^{-2} cm^{-1}), i.e. energy per unit time per unit area per unit wavelength

I_A = the integrated intensity of the absorbed radiation (erg sec^{-1} cm^{-2}), i.e. total energy absorbed per unit time per unit area

K_λ = the absorption coefficient at wavelength λ (cm^{-1})

L = the path length of uniform atomic vapour (cm)

S = the spectral bandpass of the monochromator (cm)

$I_A = \int_0^\infty I_{0\lambda}(1-e^{-K_\lambda L})\, d\lambda$ erg sec^{-1} cm^{-2}

I_0 is constant over absorption line profile

$$\therefore \qquad I_A = I_{0\lambda} \int_0^\infty (1-e^{-K_\lambda L})\, d\lambda \qquad (7)$$

At low values of $K_\lambda L$ (i.e. weak absorption), the above expression reduces to

$$I_A = I_{0\lambda} \int_0^\infty K_\lambda L \, d\lambda$$

$$= I_{0\lambda} \frac{\pi e^2 N f \lambda_0^2 L}{mc^2} \tag{8}$$

where

e = the electronic charge (CGS units)
m = the mass of the electron (g)
N = concentration of ground-state atoms (cm^{-3})
f = the oscillator strength of the absorption transition (no units)
c = the velocity of light ($cm\ sec^{-1}$).

Thus if

α = the fraction of the radiation absorbed,

$$\alpha = \frac{I_A}{I_{0\lambda} S} = \frac{\pi e^2 N f \lambda_0^2 L}{Smc^2} \tag{9}$$

Hence the calibration curves are plotted as percentage absorption versus the concentration of atoms supplied to the flame. The relationship shown in equation (9) is independent of the line broadening factors or hyperfine splitting of the absorption line. The 1 per cent absorption figures are very dependent on the monochromator bandpass and are in general somewhat poorer than those obtained using hollow-cathode lamps. It has been claimed that the stability of most continuum sources is better than that of hollow-cathode lamps, thus allowing high degrees of scale expansion to be used. It is, however, thought that modern hollow-cathode lamps are of comparable stability to continuum sources. The main disadvantage of using continuum sources is that the relationship shown in equation (9) only holds at very low percentage absorptions.

When more than 5–10 per cent of the light is absorbed (depending on the resolution of the monochromator), the calibration curve rapidly becomes convex with respect to the concentration axis (see Fig. 8.8). This is the major disadvantage of the technique, since a convex calibration curve reduces the precision of the method. It is important to bear in mind that the majority of atomic absorption analyses are performed some way above the limit of detection and that precision is usually more important than an absolute limit of detection; this is one reason why commercial manufacturers do not produce units incorporating a continuous source. The use of continuum sources has been reported by

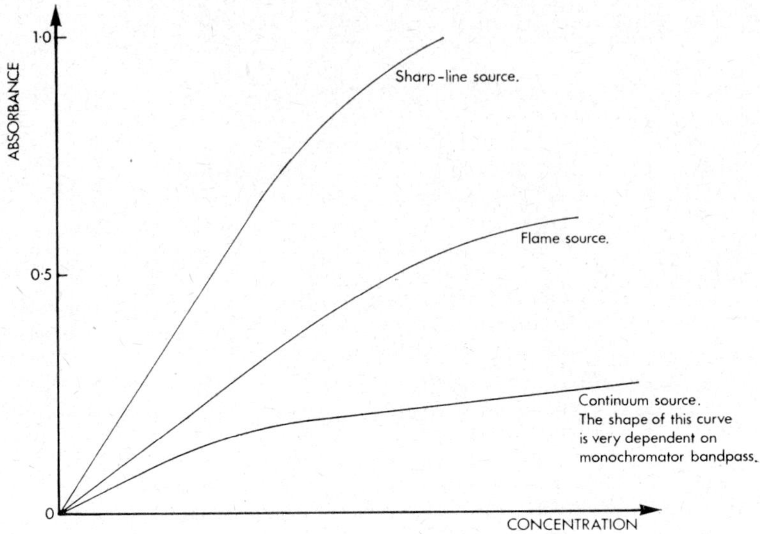

Fig. 8.8 Typical calibration curves for sharp-line, flame, and continuum source

quite a few workers(31,32,33). The main advantages and disadvantages of the use of continuous sources are shown in Table 8.3.

CASE 3

The absorption line half-width is similar to the source line half-width but less than the monochromator bandpass. In this case the source is a flame supplied with the metal to be determined, and the emission from this source flame is passed through the absorbing flame which is usually of the same type as the source flame. Thus the atoms in both flames are in identical environments and the emission and absorption line profiles will be identical. The source flame is usually shielded by an outer flame to prevent self-reversal of the resonance lines emitted. The choice of the concentration of the test element that is supplied to the source flame is a compromise between obtaining a suitably intense signal (requiring a high concentration) and minimizing self-absorption and self-reversal of the emitted radiation (requiring a low concentration). The effect of self-absorption or self-reversal on the emitted line profile is to effectively increase the half-width of the line (see Fig. 8.9).

Several workers have reported the use of flame sources(34,35,36) and RANN has given a theoretical treatment to derive the shape of the calibration curves. The theory is rather complex and makes use of a Voigt profile(4) to describe the emission- and absorption-line profiles. By

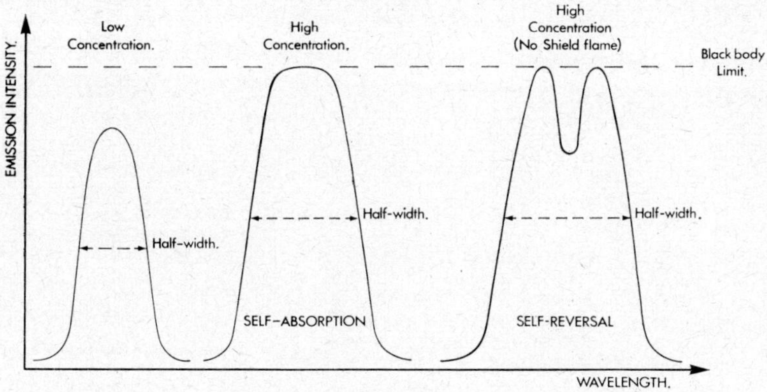

Fig. 8.9 Pictorial representation of self-absorption and self-reversal of a resonance line emitted by a flame

assuming various values of the damping constant a [$a = \sqrt{\ln 2}(\Delta\lambda_c/\Delta\lambda_D)$] and the degree of self absorption of the resonance line emitted by the source flame, absorbance versus concentration calibration curves were constructed with the aid of a computer, and the closest fit to an experimental calibration curve corresponded to an a value of 0·46 for copper, at the 3248 Å line in the air–acetylene flame.

The shape of the calibration curves are somewhat dependent on the concentration of the element supplied to the source flame (depending on the degree of self-absorption and self-reversal of the resonance line), but are generally intermediate between those of a sharp line source and those of a continuum. RANN[36] obtained sensitivities for cobalt, copper, magnesium and silver that were 30–70 per cent of those obtained using hollow-cathode lamps. The main disadvantage of using flame sources is that only relatively few metals can be readily excited to give intense enough resonance emission. Elements with lines in the far-ultra-violet, such as arsenic, cadmium and zinc give very poor emission, even in hot flames, whilst elements that form refractory oxides, such as aluminium, beryllium and molybdenum, will only emit from hot high background fuel-rich flames (e.g. nitrous oxide–acetylene) thus giving a relatively high noise-to-signal ratio.

NICKLESS and CHEICK HUSSEIN[37] have reported the use of a plasma source, which can be used to excite nearly all elements, but at the high temperature of a plasma the emitted line half-widths will tend to be wider than the absorption line profiles of atoms in flames, thus giving more convex calibration curves and poorer sensitivities than flame sources. The advantages and disadvantages of the various types of source are given in Table VIII-3.

TABLE VIII-3. COMPARISON OF TYPES OF SOURCE

Source	Advantages	Disadvantages
Hollow-cathode lamp	(1) Only one operating parameter (current) (2) Easy to stabilize (3) Moderate intensity (4) Emits very sharp lines (5) Easy to change (6) Commercially available for all elements that can be determined using atomic absorption	(1) Cannot be easily prepared (2) Expensive (3) Comparatively limited life
Microwave excited electrodeless discharge lamp	(1) Easily prepared (2) Relatively cheap (3) Long shelf life (4) Very intense, about 10–1000 times more intense than most hollow-cathode lamps (5) Emits very sharp lines	(1) Microwave generator required (2) More operating parameters than hollow-cathode lamp (cooling, position of tube within microwave cavity, and operating power) (3) At present, not commercially available for all elements that can be determined by atomic absorption
Continuum source (xenon arc lamp)	(1) Only one source required for all elements (2) Background correction for molecular absorption easily applied (3) Qualitative analysis is possible	(1) Calibration graphs only linear over a relatively short concentration range. This leads to poor precision unless all solutions are diluted to within linear range of calibration graphs. (2) Expensive, high resolution monochromator required. Sensitivity very dependent on monochromator band-pass. (3) Sensitivities usually much poorer than those obtained using sharp line sources, especially for elements with resonance lines in the far u.v. (<2500 Å) (4) Output intensity of source is low in the far u.v. (<2500Å)

TABLE VIII-3—continued

Source	Advantages	Disadvantages
Flame source	(1) Simple (2) Inexpensive (3) Useful for rare elements where only a few analyses are required (4) Multi-element sources can easily be made	(1) Sensitivities poorer than those obtained with sharp line sources (2) Stability of the emission from a flame system is not very good, hence very restricted use of scale expansion (3) Only limited number of elements will emit suitable intense resonance lines from flames (4) Non-linear calibration curves
Plasma source	(1) All elements that can be determined by atomic absorption should be excited (2) Multi-element sources can easily be made	(1) Relatively poor sensitivities compared to sharp line sources (2) Expensive high power R.F. generator required (3) Stability of the emission is not very good

Atomization in Flames

The term atomization refers to the breakdown of a compound to free atoms. When a solution is nebulized into a flame, atomization must occur before absorption of the source radiation by free atoms can occur. Most commercially available pneumatic nebulizing units have solution uptake rates of about 2–4 ml/min of which 5–15 per cent finally reaches the flame. The total flow rate of flame gases (e.g. acetylene and air or nitrous oxide) is about 8–12 litres/min. Hence it can be seen that the sample is highly diluted in the flame gases. For example, consider a nebulizer with an uptake rate of 3 ml/min, 10 per cent of which reaches the flame. Assume the total flow rate of the flame gases is 10 litres/min (ignoring air entrainment in the flame and the large expansion of the hot flame gases).

$$\text{The dilution factor} = \frac{10000}{0{\cdot}3} \approx 33000$$

Very few elements are completely atomized in the flame, thus a further effective dilution factor is involved. If the solution uptake rate is increased appreciably, the nebulized sample causes appreciable cooling of the flame gases which results in poorer atomization of the sample.

The sensitivity of atomic absorption techniques can be increased considerably by using a L'vov furnace[38], or a Far cell[39] or similar device, as an atom reservoir in place of the flame. A small volume of a sample can then be completely atomized in a relatively small volume. This has the advantage that the dilution factor associated with flames is avoided and very good sensitivities are obtained. At present these methods lack the precision and simplicity associated with flames, but with further development they should prove very useful.

Factors affecting the Percentage Atomization

The idealized processes of atomization for an indirect nebulizer, when the nebulized droplets of a solution reach the flame, are depicted in Fig. 8.10. The last two steps, volatilization and atomization, are only essentially complete for relatively few elements, e.g. Ag, Na. It would appear, for the majority of elements, that after evaporation of the water, equilibria are set up between the sample and the flame gases resulting mostly in the formation of oxide species (in a few cases hydroxide species are also formed, e.g. Ca, Cs, K). Although the sample might be nebulized as the chloride or bromide, the halides (other than perhaps certain fluorides in HF media) are usually less stable than the oxides in conventional oxygen-supported flames (in hydrogen–fluorine flames, fluorides are, in general more stable than oxides). The course of reaction can be depicted:

$$\text{Sample} \xrightarrow[\text{hydrolysis}]{\text{evaporation}} \text{clotlets} \xrightarrow{\text{volatilization}} \left| \begin{array}{l} \xrightarrow[\text{flame gases}]{\text{reaction with}} \text{MO(MOH)} \\ \qquad\qquad\qquad \downarrow\uparrow \\ \xrightarrow[\text{atomization}]{} \text{M}+\text{(O)} \end{array} \right.$$

Some oxide species have such a low vapour pressure that the oxide is present in the flame as unevaporated solid or liquid particles, e.g. Al, Mo, U, etc. This is borne out by the fact that when solutions containing aluminium are nebulized into an air–acetylene flame, irradiated with a very intense mercury lamp, weak scattering of the mercury radiation (at all the intense mercury lines) occurs. No scattering is observed when elements like zinc and cadmium, which form relatively unstable oxides, are nebulized into the same air–acetylene flame[26].

The fuel-rich air–acetylene flame has been found to give higher absorbance values for certain elements (Cr, Mg, Mo, Sn) than the hotter stoichiometric flame. This indicates that a higher degree of atomization occurs in the fuel-rich flame, and this can be attributed to the more favourable chemical environment of the fuel-rich flame although it has a lower temperature than the stoichiometric flame.

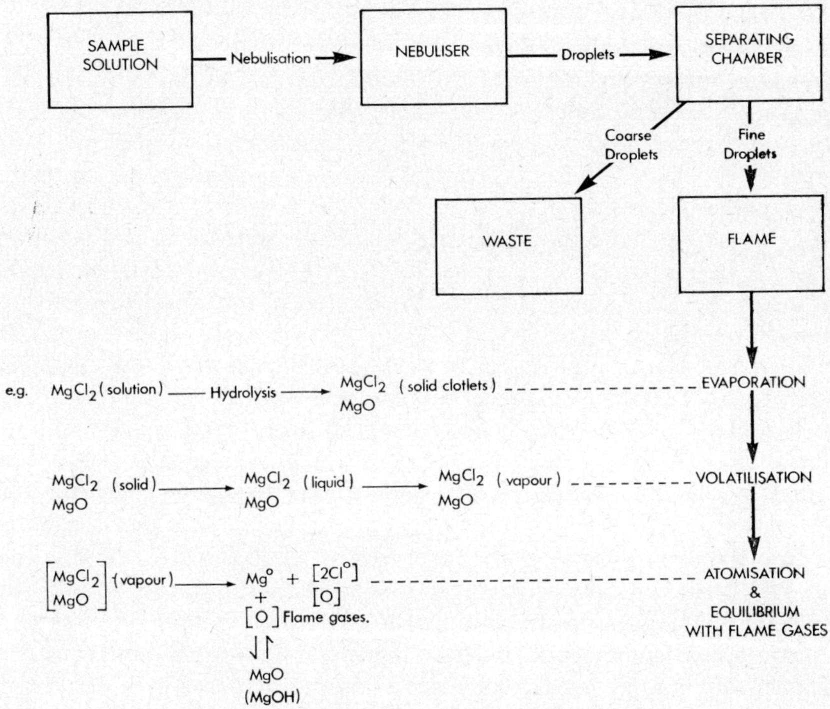

Fig. 8.10 Idealized processes of atomization

Considering the following simple dissociation reaction of a metal oxide occurring within a flame:

$$MO_{vap} \leftrightarrows M_{vap} + [O]$$

This is not meant to imply that all metals, within an oxygen supported flame, are in a simple equilibrium with their monoxides, but this simple representation can be used to explain certain observations. Assuming that the above reaction has attained a state of equilibrium, then:

$$Kp = \frac{[M]\,[O]}{[MO]}$$

$$\frac{[M]}{[MO]} = \frac{Kp}{[O]}$$

where Kp is the equilibrium constant of the above reaction. The concentration of oxygen [O] is effectively controlled by the composition and temperature of the flame gases and should be independent of any substance nebulized into the flame. It can be seen that the [M]/[MO] ratio is controlled by the value of the equilibrium constant (Kp) and the oxygen concentration [O]. The value of (Kp) will increase with increasing temperature whilst the oxygen concentration is reduced by making the flame fuel-rich. The fact that certain elements are atomized to a higher degree in a fuel-rich flame compared to a slightly hotter stoichiometric flame indicates that the small decrease in the value of (Kp) of the fuel-rich flame, compared to the hotter stoichiometric flame, is more than compensated for by a reduction in the oxygen concentration in the fuel-rich flame.

COWLEY, FASSEL and KNISELEY[40] have observed a striking enhancement in the degree of atomization of many elements in the pre-mixed fuel-rich oxygen–acetylene flame compared to the hotter stoichiometric flame. The enhancement is attributed to a favourable chemical environment provided by the interconal zone of the fuel-rich flame (for the formation and existence of free atoms).

Similar results have been observed by KIRKBRIGHT, PETERS and WEST[41] with a nitrous oxide–acetylene flame. The reducing zone of the fuel-rich flame was mainly confined to the region of red CN emission. Above the red zone of the fuel-rich flame and also in the stoichiometric flame (with no red zone) poor atomization of elements with refractory oxides, e.g. Al, Mo, Ti, etc., was observed.

The necessity of the correct chemical environment is clearly demonstrated by the fact that tin is scarcely atomized in the stoichiometric air–acetylene flame (temperature approximately 2300°C), but appreciably atomized in the very fuel-rich air–hydrogen flame and also in the cool nitrogen–hydrogen diffusion flame[42]. This is presumably due to reduction of the oxide clotlets by hydrogen in a relatively oxygen-free atmosphere.

The above discussion has assumed that the nebulized substances attain equilibrium with the flame gases. The rate of attaining equilibrium will be dependent on the nebulizer and the flame temperature. A nebulizer that produces very small droplets is better in this respect than one that produces large droplets. In general, equilibria will be established more rapidly in a hot flame than in a cool flame.

For instance, when calcium solutions containing phosphate are nebulized into an air–acetylene flame, the absorbance just above the primary reaction zone is less than that of an equilavent pure calcium solution. This decrease in the calcium free atom population is attributed to the formation of involatile calcium phosphate clotlets after evaporation of the droplets reaching the flame. These clotlets are then slowly thermally decomposed on passage through the flame thus re-establishing equilibrium between calcium atoms and the flame gases. (The interference of phosphate upon calcium almost disappears when the absorbances are measured some distance above the burner.) This type of interference can be overcome by using a hotter flame in order to decompose the involatile clotlets lower down in the flame, so that equilibrium can rapidly be set up between the metal species and the flame gases. (Phosphate has little effect on the calcium absorbance in the hot nitrous oxide–acetylene flame.)

Calcium phosphate is almost certainly formed in the clotlets, rather than by reaction in the flame gases of the air–acetylene flame, because when an air–acetylene flame is supplied by two independent nebulizers and calcium solution is supplied to one nebulizer and phosphoric acid to the other, no interference is observed. Similarly the presence of phosphorus compounds in petroleum has no effect on the calcium absorbance in the air–acetylene flame, because calcium phosphate is not readily formed in non aqueous media, and thus involatile clotlets will not be formed upon evaporation of the non aqueous droplets.

Most of the interferences observed in the air–acetylene flame that can be attributed to involatile clotlet formation can be overcome by using the hotter nitrous oxide–acetylene flame.

The determination of the percentage of the sample that is atomized in the flame is a rather complicated procedure, requiring knowledge of oscillator strengths and the absolute concentration of atoms in the

TABLE VIII-4. PERCENTAGE ATOMIZATION IN THE PRE-MIXED AIR–ACETYLENE FLAME

Element	Percentage atomization			
	Reference			
	43	8	44_{p536}	45
Ag	66			100
Al	$<10^{-3}$			
Ba	0·11	0·21	0·84	
Ca	14	4·7	8·6	
Cr	6·4	13		
Cu	98	82		37
K	25	43		
Na	52	100		40
Sr	13	11	20	
Zn	45			

flame. The published results[8,43,44,45] are not all in agreement because of the inherent errors of the measurement. The results are somewhat dependent on the burner nebulizer system used, the actual flame conditions employed, and position of measurement in the flame. Table VIII-4 shows some values that have been obtained by various workers in the fuel-rich air–aceytlene flame.

For good atomization of elements that form refractory oxides and involatile clotlet species, the hot reducing nitrous oxide–acetylene flame is to be recommended. Such elements are Al, Ba, B, Be, Ca, Ge, Mo, Si, Sr, Ti, V, W, Zr, etc. For good atomization of elements that do not form refractory oxides or involatile clotlet species, the cooler air–acetylene flame is satisfactory. Such elements are Bi, Cd, Fe, Hg, K, Na, Pb, Sb, Se, Fe, Zn, etc. The degree of atomization of these elements is not very dependent on flame conditions or flame temperature, and these elements tend to give better sensitivities in the cooler air–propane or air–coal gas flames than in the hotter air–acetylene flame.

This increase in sensitivity is probably due to the increased density of the atomic vapour at the lower temperature. The drawback of these low-temperature flames is that if the solution being nebulized has a very high solids content of elements that form refractory oxides, large clotlets will be formed. These clotlets will be more slowly decomposed in the cooler flames than in the hotter air–acetylene flame, thus the element will take longer to reach effective equilibrium with the flame gases, i.e. interference is observed.

Ionization in Flames

The population of ground-state atoms can be depleted by ionization:

$$M \leftrightharpoons M^{+} + e$$

In the air–acetylene flame very few common elements are appreciably ionized, but appreciable ionization of alkali and alkaline earth metals occurs in the hotter nitrous oxide–acetylene flame. Such elements are Na, K, Ca, Sr, Ba, etc. This type of interference is usually overcome by adding a large excess of an easily ionizable element, e.g. K, to suppress the ionization.

Isotopic Analysis by Atomic Absorption

It is possible to determine isotopic compositions using atomic emission techniques, but these usually require expensive very high resolution monochromators (resolution $\leqslant 0{\cdot}01$ Å) in order to resolve the hyperfine structure.

In theory it should be possible to determine the isotopic composition of an element by atomic absorption if certain conditions are met:

(1) Single (or at least highly enriched) isotope sources must be available for all the major isotopes of the element. The line half-widths from these sources must be somewhat less than the isotopic displacements between neighbouring components of the absorption line.

(2) The absorption line profile half-widths must be somewhat less than the isotopic displacements between neighbouring components of the absorption line.

(3) For a given isotope, the hyperfine components due to nuclear spin must themselves be partially resolved from the other isotopic components of the absorption line.

Principle of Isotopic Analysis

Consider an element with two isotopes *A* and *B*. Initially the absorption of a sample, in a suitable atom reservoir, is measured with a lamp containing pure isotope *A*. Assuming that the emission and absorption line profiles of the isotope *A* are completely resolved from those of isotope *B*, then the absorption reading is proportional to the quantity of isotope *A* in the atom reservoir. (Isotope *B* in this case will not absorb radiation emitted from isotope *A*). Similarly if this procedure is repeated using a lamp containing pure isotope *B*, the absorption reading will be proportional to the concentration of *B* in the atom reservoir. It can be seen that a high resolution monochromator is not required. Unfortunately this simple case is rarely observed in practice.

There are three main cases to be considered.

Case 1

The isotopic line absorption and emission profiles do not overlap with any other isotopic line (see Fig. 8.11). The source contains the pure isotope to be determined. This case is very rare owing to the small wavelength shifts involved. It has been observed[17] for mercury 202, which was determined using a mercury 202 cooled low-pressure microwave source and a low-pressure absorption cell, giving negligible collisional broadening (of the emission and absorption lines). The absorption cell was operated at room temperature and the Doppler width of the 2537 Å mercury absorption line under these conditions was 0·002 Å. The mercury 202 emission and absorption lines were resolved from all other isotopic lines and thus the measured absorption was directly proportional to the mercury 202 concentration. Direct determination of mercury isotopes other than mercury 200 and 202 was not possible because their absorption lines showed appreciable overlap. It is almost impossible to atomize other elements at such a low temperature, thus this case is very rare.

CASE 2

The isotopic line absorption profiles partially overlap (Fig. 8.11) and/or the source contains small quantities of other isotopes.

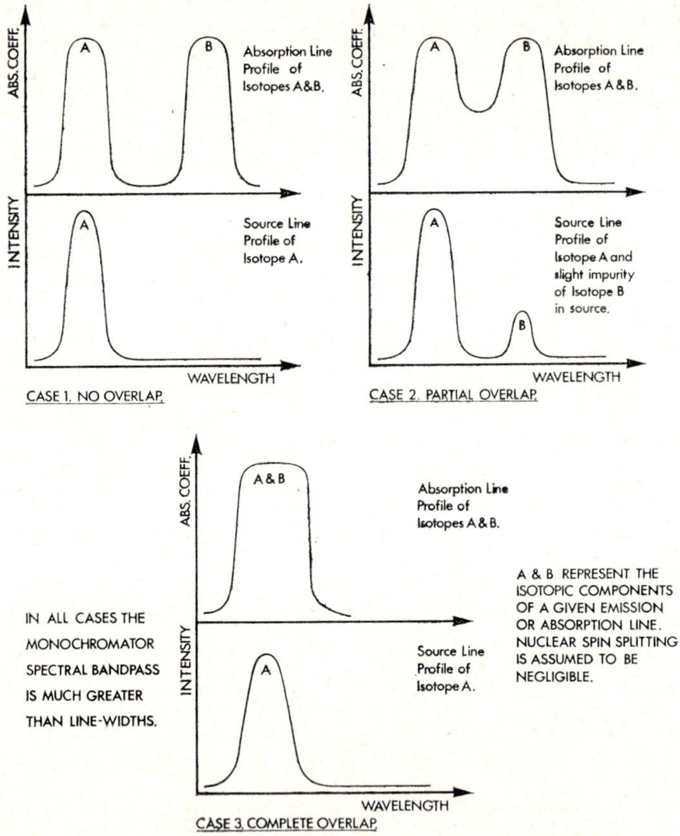

Fig. 8.11 Some possible cases of isotopic overlap of absorption line profiles for an element with two isotopes A and B

In this case part of the absorption observed using a given isotopic source is due to other isotopes. Consider a case with just two isotopes *A* and *B*. The absorption is measured using the source containing isotope *A* (or enriched isotope *A*) and then measured using the source containing isotope B (or enriched isotope B), a calibration curve then being plotted of isotopic proportion of A in the standards versus (Absorbance using source A/Absorbance using source B).

This situation has been observed in the case of lithium, the 6708 Å resonance line is a doublet (0·16 Å separation) each line of which is split into isotopic components also 0·16 Å apart (see Fig. 8.12). (The nuclear spin splitting is about 0·01 Å, see Table 8.1.) The determination of lithium isotopes has been reported using hollow-cathode lamps[22], flame sources[16] of lithium 6 and lithium 7, and a flame as an absorption cell. (The absorption and emission line half-width of the 6708 Å lines in the air–acetylene flame is reported as 0·13 Å[24].) Using flames there is appreciable overlap of the isotopic components but it was possible to determine lithium isotopes using a flame as the absorption cell[16,22].

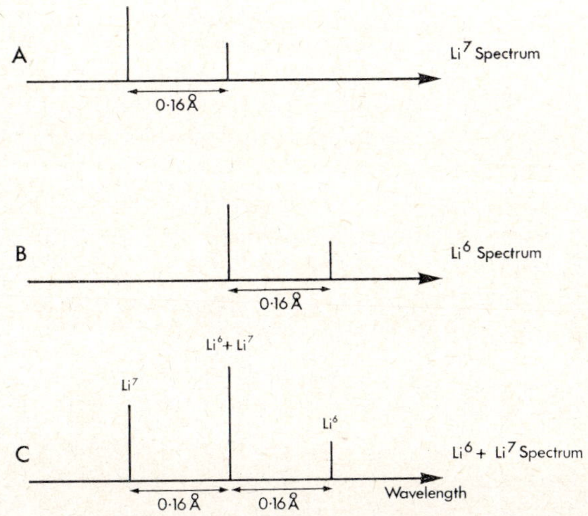

Fig. 8.12 Isotopic splitting of the lithium 6708 Å doublet

A more accurate measurement was made by GOLEB and YOKOYAMA[21] who used two water-cooled Schuler–Gollnow lithium hollow-cathode lamps, one containing lithium 6 the other containing lithium 7. A modified water-cooled Schuler–Gollnow hollow-cathode tube was used to vaporize the samples and standards and acted as the absorption cell. Both emission and absorption tubes were at low pressures in order to minimize collisional broadening.

Similarly GOLEB has determined uranium 235 and 238[20] by a similar technique (using water-cooled Schuler–Gollnow tubes). In this case the 4153 and 5027 Å uranium lines were used, as these lines of uranium 235 have negligible nuclear spin splitting but exhibit an appreciable isotope shift between the 235 and 238 isotopes of 0·07 and 0·10 Å respectively. The sources contained the enriched isotopes, not the pure 235 and 238 isotopes. GOLEB also demonstrated a variation on the above method[20] whereby instead of placing samples and standards in the absorption tube, one source lamp contained the sample and standards (one at a time) whilst the absorption tube contained the pure isotope (or the enriched isotope) to be determined. In this case a linear calibration curve was plotted between percentage of uranium 238 in the standards, versus the percentage transmission observed through the absorption cell. The advantages of this technique are that metal chips, oxides, and compounds can be used directly as samples and that the proportion of incident light absorbed is independent of the emission intensity (whilst when samples are placed in the absorption tube, the absorbance is very dependent on spluttering current flowing in the absorption tube.)

Little overlap of the uranium isotopic components was observed. This was partially due to the large atomic mass of uranium, as the line widths are limited by Doppler broadening and the half-width of a line is inversely proportional to the square root of the atomic weight of an element. Thus for a given Doppler temperature the uranium line half-width will be $\sqrt{(235/7)}$ times smaller than the lithium line half-width (ignoring the difference between the wavelengths of the lines).

Similarly KIRCHHOFF[46] has determined the lead 206 and lead 208 isotopic constitution of lead samples using the 2833 Å line. In this case the samples and standards are excited in a hollow-cathode lamp and the radiation is passed through a modified hollow-cathode lamp containing pure lead 208 (or pure lead 206). The calibration curve is then constructed by plotting the percentage transmission against the percentage of lead 208 (or lead 206) in the standards. The 2833 Å lead line has four main hyperfine components (ignoring the small contribution of Pb204), see Fig. 8.13. It can be seen from the figure that isotopic splitting of the 206 and 208 components is much less than the nuclear spin splitting of the 207 component. (The 206 and 208 components exhibit no nuclear spin splitting.)

Thus it is important to realize that for the determination of lithium, uranium and certain other elements in flame cells, by atomic absorption, the element in the source lamp must have the same isotopic constitution as the element in the samples and standards.

CASE 3

The isotopic line absorption profiles completely overlap. GOLEB[23] has observed this case when trying to determine boron 10 and 11 using an absorption tube. The isotopic shift of both lines of the 2496/7 Å doublet is only 0·01 Å and with a given boron isotopic lamp, the same absorption figures were obtained when either boron 10 or boron 11 was in the absorption tube.

It would appear strange that it is possible to determine lead 206 and lead 208 (isotopic splitting 0·007 Å) and not boron 10 and boron 11 (isotopic splitting 0·01 Å). The line widths are controlled by Doppler broadening, and the Doppler half-width of an absorption or an emission line is proportional to the square root of the temperature, and inversely proportional to the square root of the atomic weight of the absorbing or emitting atoms. Boron is far more refractory than lead and therefore will require a much higher temperature, i.e. spluttering current, to produce an appreciable atom population. Also, from atomic weight considerations, the boron line, for a given temperature, will be $\sqrt{(207/11)}$ times broader than the lead line if the relatively small difference between the wavelengths of the two lines is ignored.

It can be seen that although isotopic analysis can be performed using atomic absorption techniques, it is a more complicated procedure than conventional atomic absorption techniques. Many isotope shifts (Table VIII-2) are somewhat less than average absorption line half-widths, in conventional flames, thus flames are not really suitable as absorption cells, although the use of low-pressure flames which will minimize collisional broadening should be more feasible. Up to the present only two isotopes have been determined for a given element. For an element with more than two major isotopes (assuming partial overlap) the procedure becomes more complicated. The determination of minor

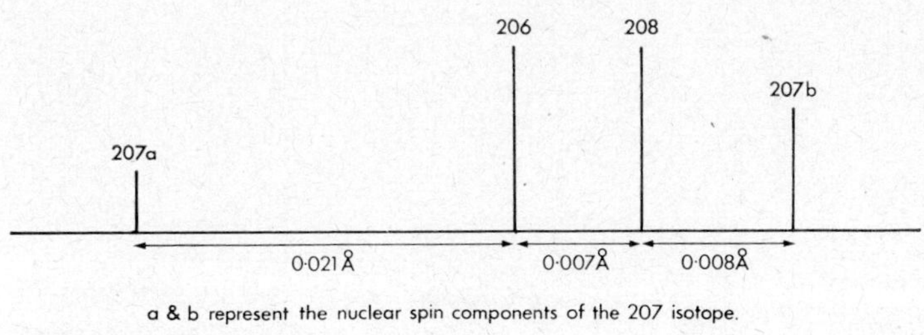

Fig. 8.13 Hyperfine splitting of the 2833 Å lead line

isotopic constituents (< 2 per cent) is thought to be impractical. Thus isotopic analysis using atomic absorption is thought to have limited application.

Comparison of Atomic Absorption and Flame Emission Spectroscopy

Consider an atomic vapour at a given temperature in a uniform media (a flame supplied with a given species from a pneumatic nebulizer is a fair approximation to this). The ratio of the number of atoms in the excited state (N_1) to the number in the ground state (N_0) is given by the Boltzmann equation:

$$\frac{N_1}{N_0} = \frac{g_1}{g_0}\exp\left(\frac{-E}{kT}\right)$$

where

g_0 and g_1 = the statistical weights of the ground and excited states respectively
k = Boltzmann constant (erg °K^{-1})
T = absolute temperature (°K)
E = energy of the excited state above the ground state (erg)

Except for elements with long-wavelength resonance lines (K, Cs) at extremely high temperatures, the number of atoms in the lowest excited state is negligible in comparison to the number in the gound state (see Table VIII-5).

TABLE VIII-5. RATIO OF THE NUMBER OF ATOMS IN THE EXCITED STATE TO THE NUMBER IN THE GROUND STATE[44]

Resonance line (Å)	g_1/g_0	Excitation energy (e.v.)	N_1/N_0		
			T = 2000	2500	3000°K
Na, 5890	2	2·10	$0{\cdot}99\times10^{-5}$	$1{\cdot}14\times10^{-4}$	$5{\cdot}83\times10^{-4}$
Ba, 5536	3	2·24	$6{\cdot}83\times10^{-6}$	$3{\cdot}19\times10^{-5}$	$5{\cdot}19\times10^{-4}$
Ca, 4227	3	2·93	$1{\cdot}22\times10^{-7}$	$3{\cdot}67\times10^{-6}$	$3{\cdot}55\times10^{-5}$
Cu, 3248	2	3·81	$4{\cdot}82\times10^{-10}$	$4{\cdot}04\times10^{-8}$	$6{\cdot}65\times10^{-7}$
Mg, 2852	3	4·35	$3{\cdot}35\times10^{-11}$	$5{\cdot}20\times10^{-9}$	$1{\cdot}50\times10^{-7}$
Zn, 2139	3	5·80	$7{\cdot}45\times10^{-15}$	$6{\cdot}22\times10^{-12}$	$5{\cdot}50\times10^{-10}$

The ratio N_1/N_0 increases exponentially with temperature and wavelength. Thus it can be seen that at flame temperatures of 2–3000°C effectively 100 per cent of the atoms are in the ground state and that the ratio (N_1/N_0) rapidly decreases with decreasing wavelength (increasing E_1). Thus flame emission methods are not very sensitive for elements with resonance lines below 2700 Å, e.g. As, Bi, Cd, Hg, Sb, Se, Te, Zn, etc. Atomic absorption methods are only dependent on the ground-state atom population and not on the wavelength of the resonance line.

The number of excited atoms varies exponentially with temperature (see Table VIII-5) whilst the number of ground-state atoms remains essentially constant. Thus any slight temperature fluctuation in an atomic vapour will have a very pronounced effect on the excited state atom population used in flame emission, and hardly affect the ground state atom population used in atomic absorption. An increase in temperature will cause a small decrease in the density of the atomic vapour, and will slightly alter the degree of atomization.

Spectral interference is often observed in flame emission owing to the large number of atomic lines and broad molecular bands (e.g. CaOH, MgO, CuOH) that can be observed in emission. In atomic absorption studies, the primary radiation source is modulated either electronically or mechanically and an a.c. amplifier, tuned to the modulation frequency, is connected to the detector. Thus any atomic or molecular emission from the flame should not be detected by the measuring system.

Most elements have simple atomic absorption spectra consisting of few lines, and spectral interference between absorption lines of different elements is seldom observed. Molecular absorption bands in hot flames are usually weak (low transition probabilities) and can generally be ignored unless very high concentrations of substances (>5000 μg/ml) are being nebulized.

It is often assumed that flame emission techniques are more prone to inter-element effects than atomic absorption techniques. Although this is true of spectral interference it must be realized that any factor that affects the ground-state atom population, e.g. ionization, compound formation in the flame, will affect absorption and emission techniques equally.

Comparison of Atomic Absorption and Atomic Fluorescence Spectroscopy

The principles of atomic fluorescence spectroscopy have already been discussed. Atomic fluorescence is essentially an emission technique, i.e. an absolute intensity is measured, whilst atomic absorption is a comparative technique, i.e. the ratio of two intensities are measured. The respective merits and demerits of the two techniques are summarized below.

A F S—ADVANTAGES IN RELATION TO A A S

1. A F S is essentially an emission technique and the sensitivity can be increased by either increasing the source intensity or the gain of the instrument. In A A S the absorbance is a ratio ($\log_{10} I_0/I_T$), hence any increase in the source intensity I_0 is accompanied by a corresponding increase in I_T. Hence the ratio $\log_{10} I_0/I_T$ remains unaltered.

2. A F S requires somewhat less sophisticated apparatus and burners than A A S. Optical alignment of the various components is usually simpler.

3. A continuous source can be used in A F S with a normal monochromator (the detector does not 'see' the source), whilst the use of a continuous source in A A S requires the use of a high resolution monochromator.

4. In general, for most elements, there are more fluorescence lines than absorption lines. This gives a greater choice in the wavelength of measurement, e.g. the main arsenic resonance lines are at 1890, 1937, and 1972 Å where absorption by optics and flame gases can be quite serious, and also the photomultiplier response is often poor at these wavelengths, especially for the most sensitive 1890 Å line. Although these resonance lines must be used in absorption measurements (often the 1890 Å line cannot even be detected), the fluorescence resulting from the absorption of the 1890 Å line can be measured at 2288 or 2350 Å[47].

5. The design of an instrument that uses filters in place of a monochromator is much easier for A F S than A A S because the detector does not 'see' the source. Also the use of a filter enables the detector to be brought much nearer to the flame, thus increasing the sensitivity of A F S.

A F S—DISADVANTAGES IN RELATION TO A A S

1. Atomic absorption measurements are independent of the quantum efficiency but fluorescence measurements are directly proportional to the quantum efficiency. As long as the flame temperature and composition are kept constant and the partial pressures of the nebulized species are low, which is generally the case, inter-element effects on the quantum efficiency should be negligible.

2. The determination of refractory metals by A F S using conventional nitrous oxide–acetylene flames is not very satisfactory. This is because A F S is essentially an emission technique and larger slit widths and higher gains than are commonly employed in A A S are necessary at low concentrations. Although the light from the source can be modulated so that the high intensity background radiation from the

flame is not amplified, this background can nevertheless seriously overload the photomultiplier and increase the noise level.

3. Light scattering by particles, e.g. refractory oxides, in the flame is seldom observed in A A S. In fact using conventional instruments it is difficult to detect 0·1 per cent scatter of the incident radiation.

Light scatter is a greater potential source of interference in A F S than in A A S. This is because the integrated intensity in all directions of a given fluorescence line at a given wavelength is seldom greater than 0·2 per cent of the integrated intensity of the incident line from the source reaching the flame at the same wavelength(26). At low concentrations the figure is much lower. Thus 0·1 per cent scatter of the incident radiation represents a far greater source of interference in A F S than in A A S. Source scattering is far more serious when a continuous source is used in A F S because the integrated intensity of the scatter signal reaching the detector is then dependent on the square of the monochromator band-pass.

However, with a well designed nebulizer-burner unit and using line sources, source scattering is seldom observed in A F S.

4. The sensitivity of atomic absorption is almost independent of the source intensity. If only a low intensity source is available atomic fluorescence is not very sensitive, because the sensitivity of this technique is proportional to the source intensity.

Chap. VIII References

1. WOLLASTON, W. H., *Phil. Trans.* 92 (1802) 365.
2. FRAUNHOFER, J., *Ann. d. Physik* 56 (1817) 264.
3. KIRCHHOFF, G. R., *Ann. d. Physik* 109 (1860) 148, 275. *Phil. Mag.* 20 (1860) 1.
4. MITCHELL, A. C. G., and ZEMANSKY, M. W., *Resonance Radiation and Excited Atoms* (University Press, Cambridge, 1961).
5. WALSH, A., The application of atomic absorption spectra to chemical analysis. Australian Patent No. 23041/53. *Spectrochim. Acta* 7 (1955) 108.
6. PARSONS, M. L., MCCARTHY, W. J., and WINEFORDNER, J. D., Approximate half-intensity widths of a number of atomic spectral lines used in atomic-emission and atomic-absorption flame spectroscopy. *Appl. Spec.* 20 (1966) 223.
7. HINNOV, E., A method of determining optical cross sections. *J. Opt. Soc. Amer.* 47 (1957) 151.

8. HINNOV, E., and KOHN, H., Optical cross sections from intensity–density measurements. *J. Opt. Soc. Amer.* 47 (1957) 156.

9. YASUDA, K., Relationship between resonance line profile and absorbance in atomic absorption spectroscopy. *Anal. Chem.* 38 (1966) 592.

10. ALKEMADE, C. Th. J., Science vs. fiction in atomic absorption. *Appl. Opt.* 7 (1968) 1261.

11. CANDLER, C., *Atomic Spectra* (Hilger and Watts, 2nd Ed., London, 1964).

12. KUHN, H. G., *Atomic Spectra* (Longmans, London, 1964).

13. COX, A. H., Wavelengths in the spectrum of ^{86}Kr (1) between 6701 and 4185 Å. *J. Opt. Soc. Amer.* 55 (1965) 780.

14. MEGGERS, W. F., and WESTFALL, F. O., Lamps and wavelengths of mercury 198. *J. Res. Nat. Bur. Stds.* 44 (1950) 447.

15. DAVISON, A., GIACCHETTI, A., and STANLEY, R. W., Interferometric wavelengths of thorium lines between 2650 Å and 3400 Å. *J. Opt. Soc. Amer.* 52 (1962) 447.

16. MANNING, D. C., and SLAVIN, W., Lithium isotope analysis by atomic absorption spectrophotometry. *At. Abs. Newlsetter*, No. 1 (1962) 39.

17. OSBORN, K. R., and GUNNING, H. E., Determination of Hg^{202} and other mercury isotopes in samples of mercury vapour by mercury resonance radiation absorbiometry. *J. Opt. Soc. Amer.* 45 (1955) 552.

18. ATKINSON, R. J., CHAPMAN, G. D., and KRAUSE, L., Light sources for the excitation of atomic resonance fluorescence in potassium and rubidium. *J. Opt. Soc. Amer.* 55 (1965) 1269.

19. BURLING, D. H., CZAJKOWSKI, M., and KRAUSE, L., Light sources for the excitation of atomic resonance fluorescence in caesium and sodium. *J. Opt. Soc. Amer.* 57 (1967) 1162.

20. GOLEB, J. A., The determination of uranium isotopes by atomic absorption spectrophotometry. *Anal. Chim. Acta* 34 (1966) 135.

21. GOLEB, J. A., and YOKOYAMA, Y., The use of a discharge tube as an absorption source for the determination of Lithium-6 and Lithium-7 isotopes by atomic absorption spectrophotometry. *Anal. Chim. Acta* 30 (1964) 213.

22. ZAIDEL, A. N., and KORENNOI, E. P., Spectral determination of the isotopic composition and concentration of lithium in solution. *Optics and Spectroscopy* 10 (1961) 299.

23. GOLEB, J. A., An attempt to determine the boron natural abundance ratio B^{11}/B^{10} by atomic absorption spectrophotometry. *Anal. Chim. Acta* 36 (1966) 130.

24. SOBELEV, N. N., The shape and width of spectral lines emitted by a flame and a d.c. arc. *Spectrochim. Acta* 11 (1957) 310.

25. DAGNALL, R. M., THOMPSON, K. C., and WEST, T. S., The fluorescence characteristics and analytical determination of bismuth with an iodine electrodeless discharge tube as source. *Talanta* 14 (1967) 1467.

26. THOMPSON, K. C., and WILDY, P. C., Electrically modulated microwave excited discharge tubes in atomic spectrocopy. *Analyst*, submitted for publication.

27. FASSEL, V. A., RAMUSON, J. O., and COWLEY, T. G., Spectral line interferences in atomic absorption spectroscopy. *Spectrochim. Acta* 23B (1968) 579.

28. MANNING, D. C., and FERNANDEZ, F., Cobalt spectral interference in the determination of mercury. *At. Abs. Newsletter* 7 (1968) 24.

29. ALLAN, J. E., A spectral interference in atomic absorption spectroscopy. *Spectrochim. Acta* 24B (1969) 13.

30. ZEEGERS, P. J. T., SMITH, R., and WINEFORDNER, J. D., Shapes of analytical curves in flame spectrometry. *Anal. Chem.* 40 (1968) 26A.

31. FASSEL, V. A., MOSSOTTI, V. G., GROSSMAN, W. E. L., and KNISELEY, R. N., Evaluation of spectral continua as primary sources in atomic absorption spectroscopy. *Spectrochim. Acta* 22 (1966) 347.

32. MCGEE, W. W., and WINEFORDNER, J. D., Use of a continuous source of excitation, and argon–hydrogen–air flame and an extended flame cell for atomic absorption flame spectrophotometry. *Anal. Chim. Acta* 37 (1967) 429.

33. ALLAN, J. E., Atomic absorption spectrophotometry absorption lines and detection limits in the air–acetylene flame. *Spectrochim. Acta* 18 (1962) 259.

34. SKOGERBOE, R. K., and WOODRIFF, R. A., Atomic-absorption spectra of europium, thulium and ytterbium using a flame as a line source. *Anal. Chem.* 35 (1963) 1977.

35. ALKEMADE C. T. J., and MILATZ, J. M. W., A double beam method of spectral selection with flames. *J. Opt. Soc. Amer.* 45 (1955) 583.

36. RANN, C. S., Evaluation of a flame as the spectral source in atomic absorption spectroscopy. *Spectrochim. Acta* 23B (1968) 245.

37. CHEIKH HUSSEIN, A. M., and NICKLESS, G., An investigation into the R. F. plasma as an excitation source in atomic absorption and fluorescence spectrometry. (International Atomic Absorption Spectroscopy Conference, Sheffield, July 1969).

38. L'VOV, B. V., The analytical use of atomic absorption spectra. *Spectrochim. Acta* 17 (1961) 761.

39. WEST, T. S., and WILLIAMS, X. K., Atomic absorption and fluorescence spectroscopy with a carbon filament atom reservoir. Part I. Construction and operation of atom reservoir. *Anal. Chim. Acta* 45 (1969) 27.

40. COWLEY, T. G., FASSEL, V. A., and KNISELEY, R. N., Free atom formation processes in premixed fuel-rich and stoichiometric oxygen–acetylene flames employed in atomic emission and absorption spectroscopy. *Spectrochim. Acta* 23B (1968) 771.

41. KIRKBRIGHT, G. F., PETERS, M. K., and WEST, T. S., Emission spectra of nitrous oxide supported acetylene flames at atmospheric pressure. *Talanta* 14 (1967) 789.

42. THOMPSON, K. C., Unpublished studies.

43. DE GALAN L., WINEFORDNER, J. D., Measurement of the free atom fraction of 22 elements in the air–acetylene flame. *J. Quant Spectrosc. Radiat. Transfer* 7 (1967) 251.

44. MAVRODINEANU, R., and BOITEUX, H., *Flame Spectroscopy* (Wiley, New York, 1965).

45. RANN, C. S., Absolute analysis by atomic absorption. *Spectrochim. Acta* 23B (1968) 827.

46. KIRCHHOFF, H., Bestimmung des Isotopenmischungsverhältnisses von Bleiproben mit der Atomabsorptionsmethode. *Spectrochim. Acta* 24B (1969) 235.

47. DAGNALL, R. M., THOMPSON, K. C., and WEST, T. S., Fluorescence and analytical characteristics of arsenic, with a microwave. excited electrodeless discharge tube as source. *Talanta* 15 (1968) 677.

Index